Wireless Networks

The purpose of Springer's Wireless Networks book series is to establish the state of the art and set the course for future research and development in wireless communication networks. The scope of this series includes not only all aspects of wireless networks (including cellular networks, WiFi, sensor networks, and vehicular networks), but related areas such as cloud computing and big data. The series serves as a central source of references for wireless networks research and development. It aims to publish thorough and cohesive overviews on specific topics in wireless networks, as well as works that are larger in scope than survey articles and that contain more detailed background information. The series also provides coverage of advanced and timely topics worthy of monographs, contributed volumes, textbooks and handbooks.

** Indexing: Wireless Networks is indexed in EBSCO databases and DPLB **

Jiannong Cao • Yanni Yang

Wireless Sensing

Principles, Techniques and Applications

Jiannong Cao
Hong Kong Polytechnic University
Hong Kong, China

Yanni Yang
Hong Kong Polytechnic University
Kowloon, China

ISSN 2366-1186 ISSN 2366-1445 (electronic)
Wireless Networks
ISBN 978-3-031-08347-1 ISBN 978-3-031-08345-7 (eBook)
https://doi.org/10.1007/978-3-031-08345-7

This Springer imprint is published by the registered company Springer Nature Switzerland AG
The registered company address is: Gewerbestrasse 11, 6330 Cham, Switzerland

Preface

Motivation of This Book

Wireless signals play a crucial role in communication for people's daily lives. Without wireless signals, for example, Wi-Fi, Bluetooth, and radio-frequency identification (RFID) signals, people may not enjoy those convenient and joyful services. In the last decade, apart from the communication function, researchers have shown that wireless signals provide a new paradigm for sensing human activities and objects. Wireless sensing has many benefits and advantages over traditional sensing approaches. First, wireless devices are prevalent almost everywhere. Using existing wireless infrastructures can provide a ubiquitous and cost-effective way for sensing. Second, wireless sensing can be non-intrusive. People can get rid of the traditional bulky sensors and avoid being monitored by the camera.

The primary motivation of this book is to provide a comprehensive summary of common principles of wireless sensing, a systematic study about challenging issues in complex and practical scenarios, and critical solutions to tackle these challenges. In particular, we first present a systematic introduction for the development and the vision of wireless sensing technology. Second, we provide a comprehensive study about wireless signals, sensing principles, and applications. Third, we discuss the challenges of putting wireless sensing into tangible and practical use. Then, we introduce methods to tackle the challenging issues in wireless sensing. Finally, we share our experiences in designing wireless sensing systems for several essential applications.

What This Book Is About

This book provides comprehensive reviews of the wireless sensing technology, including the types of wireless signals used for sensing, the general principles and approaches for wireless sensing, the information that wireless signals can sense, and the key applications. This book presents an investigation of the critical challenges that need to be alleviated to achieve wireless sensing in complex and practical scenarios and how to tackle these challenges with concrete case studies and examples. The book can be seen as a textbook and a practical guide for the reader.

How This Book Is Organized

This book is divided into six chapters:

Chapter 1: "Introduction". This chapter introduces motivations, basics, applications, and the framework of wireless sensing.

Chapter 2: "Wireless Signals and Signal Processing". This chapter introduces different kinds of wireless signals and signal processing methods utilized for wireless sensing.

Chapter 3: "Wireless Sensing System Configurations". This chapter introduces and compares different configurations of wireless devices in wireless sensing systems.

Chapter 4: "Wireless Sensing Methodologies". This chapter introduces and discusses representative methodologies to obtain sensing information from wireless signals.

Chapter 5: "Case Studies". This chapter showcases representative case studies of applying wireless sensing for different applications, including human respiration monitoring, exercise monitoring, and liquid sensing.

Chapter 6: "Conclusion". This chapter provides the summary and future directions of wireless sensing.

Hong Kong, China Jiannong Cao
Kowloon, China Yanni Yang
March 2022

Acknowledgments

The authors are deeply grateful to the research staff and students in our research group for their hard work in developing wireless sensing systems and applications. We express our thanks to Dr. Xuefeng Liu, Dr. Xiulong Liu, Dr. Zhuo Li, Dr. Yanwen Wang, Dr. Shan Jiang, Mr. Jiuwu Zhang, and Mr. Jinlin Chen. The financial support from the Hong Kong RGC Research Impact Fund (R5034-18) and Collaborative Research Fund (CRF)—Group Research Grant (C5026-18G) is greatly appreciated.

Contents

List of Symbols and Abbreviations

a	Signal amplitude of a single path
A	Amplitude of overall received signal
B	Bandwidth
c	Light speed
d	Signal propagation distance
Δd	Change of signal propagation distance
f	Frequency
$h(t)$	Impulse response
$H(f)$	Frequency-domain representation of $h(t)$
v	Signal speed
$x(t)$	Transmitted signal
$X(f)$	Frequency-domain representation of $x(t)$
$y(t)$	Received signal
$Y(f)$	Frequency-domain representation of $y(t)$
φ	Signal phase
$\Delta\varphi$	Phase difference
ϕ	Azimuth angle of arrival
θ	Elevation angle of arrival
λ	Signal wavelength
τ	Time delay
$*$	Convolution operation
ADC	Analog-to-digital converter
AM	Amplitude modulation
AOA	Angle of arrival
AP	Access point
BSS	Blind source separation
CFO	Carrier frequency offset
CFR	Channel frequency response
CSI	Channel state information
CSS	Chirp spread spectrum
CW	Continuous wave

DAC	Digital-to-analog converter
DWT	Discrete wavelet transform
FFT	Fast Fourier transformation
FM	Frequency modulation
FMCW	Frequency-modulated carrier wave
HPA	High power amplifier
ICA	Independent component analysis
IFFT	Inverse fast Fourier transformation
IoT	Internet of Things
IQ	In-phase and quadrature
ISAR	Inverse synthetic aperture radar
LFM	Linear frequency modulation
LNA	Low noise amplifier
LOS	Line of sight
LPF	Low-pass filter
LTI	Linear time-invariant
MIMO	Multiple-input multiple-output
MUSIC	Multiple signal classification
OFDM	Orthogonal frequency-division multiplexing
PA	Power amplifier
PDD	Packet detection delay
PGA	Programmable gain amplifier
PPO	Phase-locked loop
RF	Radio frequency
RFID	Radio frequency identification
RSS	Received signal strength
Rx	Receiver
SFO	Sampling frequency offset
SNR	Signal-to-noise ratio
STFT	Short-term Fourier transformation
SVM	Support vector machine
TOF	Time of flight
Tx	Transmitter
UHF	Ultra-high frequency
UWB	Ultra-wide band

Chapter 1
Introduction

1.1 Wireless Sensing and Applications

The definition of *sensing* is very broad. People in different fields have their own understanding about the scope of sensing technologies. A prevailing view is that sensing technologies require sensors to acquire information of the sensing target prior to converting the information into a readable output. In today's smart and connected society, we have witnessed the fast-growing and population of sensors. There are two mainstream sensors: vision-based sensors (e.g., video cameras and infrared sensors) and wearable sensors (e.g., smartwatch and glasses). Vision-based sensors can directly capture the image of the target. However, vision-based sensing can bring many privacy concerns, especially for sensing human beings. The abuse of cameras makes people to be nervous and sensitive to vision-based sensing solutions. Wearable sensors are less privacy-intruding than cameras. The sensing target needs to wear the sensor to measure the target's physical, chemical, and biological reactions via direct contact. However, many wearable sensors are heavy and expensive. For human behavior sensing, it would be quite cumbersome to wear heavy or tight sensors on the human body. For sensing a large number of objects, attaching sensors to each object is costly.

Considering the above concerns of existing sensing technologies, in recent decades, researchers have proposed a new sensing paradigm called *wireless sensing*. The presence of wireless sensing stems from the rapid prevalence and pervasiveness of wireless communication technologies in people's everyday lives. Researchers find that wireless signals traditionally used for communication can also serve for sensing tasks. In a word, wireless sensing is a technology that can sense human beings, objects, and surrounding environments by using wireless signals. Different kinds of wireless signals, e.g., acoustic signals, radio frequency (RF) signals, and light signals, have been investigated and vastly applied for various kinds of sensing applications [1–3].

J. Cao, Y. Yang, *Wireless Sensing*, Wireless Networks,
https://doi.org/10.1007/978-3-031-08345-7_1

Wireless sensing has many overwhelming benefits and advantages. First, wireless signals are pervasively used everywhere in people's daily lives. Nowadays, most indoor places are equipped with WiFi routers. Cellular networks (e.g., the LTE network) have covered over 80% of the world population. Bluetooth, WiFi, acoustic, and light sensors can be embedded in a single smartphone or laptop. Hence, we can leverage existing wireless infrastructure and cost-effective wireless devices to save extra expenses when developing or purchasing new sensing equipment. Second, wireless signals enable non-intrusive human sensing compared with existing human sensing approaches, which rely on cameras or wearable sensors. When using cameras, human appearance can be captured, resulting in privacy intrusion of the sensed person. Besides, sensing accuracy is negatively affected in poor lighting conditions. Wearing sensors on the body, especially for bulky and tight sensors like the headset and chest belt, brings users much more inconvenience. By using wireless signals, privacy concerns can be eliminated, and human beings can even be sensed in the dark. Wireless signals can also achieve human sensing without requiring users to wear any sensor. Finally, wireless signals also enable battery-free sensing, which is demanded by large-scale structures or equipment monitoring. Many wireless tags, e.g., radio frequency-identification (RFID) and WiFi tags, can be powered up remotely by electromagnetic signals. By attaching lightweight wireless tags to the target, it can be sensed in the long term without changing the battery from time to time.

The general principles of wireless sensing can be viewed in twofold. First, for motion-related sensing, movements of the target, either the human or object, can affect the propagation of wireless signals. For example, when the target carries a wireless transmitter and moves away from the wireless receiver, the length of the signal propagation path between the transmitter and receiver will increase and more signal power will be attenuated. Therefore, we can extract many motion-related information, including moving speed, frequency, distance, and direction, from the wireless signal indicators. We can further derive the semantic information of motions for activity recognition. Second, for sensing static objects or environments, the presence of an object and the surrounding environment can change the medium (also called "channel") in which wireless signals are propagating. For example, a higher environment temperature can increase the propagation speed of acoustic signals. Therefore, we can detect the presence or identify the object and extract environmental information from wireless signals. Various kinds of wireless signals have been exploited for different sensing purposes. Different wireless signals have distinct characteristics, e.g., the wavelength, frequency band, and configurations. The selection of the wireless signal requires careful considerations of application requirements, such as the scope of target users, availability of wireless devices, and expected sensing performance.

Wireless sensing has been adopted in many critical applications, e.g., healthcare and assisted living, security and surveillance, human-computer interaction, indoor navigation, and industrial automation. We select and introduce some representative wireless sensing applications.

1.1.1 Applications in Healthcare and Assisted Living

The current aging society and sub-health issues call for smart healthcare technologies that can be conveniently applied in people's daily lives. Wireless sensing has promoted different aspects of healthcare and assisted living applications, including vital sign monitoring, exercise monitoring, and daily activity recognition.

Vital Sign Monitoring Three vital signs have been monitored using wireless signals, including respiration rate [4–6], heartbeat [7, 8], and blood pressure [9, 10]. The chest movement during respiration has a large displacement compared with the heartbeat. Zigbee, WiFi, RFID, LoRa, radar, and acoustic signals can all be used to detect the respiration activity. Using WiFi and LoRa signals can even achieve respiration monitoring when the target person is behind the wall of the signal transmitter (Tx) and receiver (Rx) [11]. Meanwhile, the monitoring of heartbeat needs the wireless signal sensor to be directly attached to the chest (e.g., RFID tag) or high-resolution signal measurements for contact-free systems (e.g., FMCW radar). Apart from respiration and heart rate, sleep apnea and inter-beat-interval (IBI), which are related to respiration and heartbeat, can also be detected. For example, researchers attached RFID tags on the human chest to measure the IBI from the RFID signal [7]. Doppler radar is used to estimate the beat-to-beat blood pressure from the pulse transmit time. The radar is placed around the sternum to acquire the aortic arterial pulsation.

Exercise Monitoring Wireless signals have been used for different exercise activities, e.g., running, cycling, and free-weight exercise activities. Specifically, the CSI amplitude of the WiFi signal is used to count the running step of people on the treadmill by detecting rhythmic running patterns [12]. RFID tags are attached to fitness instruments, e.g., dumbbell and barbell, to recognize, count, and assess free-weight activities [13]. Extracting the Doppler profile of each activity from the RFID signal phase can help to distinguish different free-weight activities. RFID signal also enables the simultaneous monitoring of respiration and exercise activities and estimate the locomotion respiration coupling ratio [14]. Apart from recognizing and counting exercise activities, wireless signals are also employed for personalized assistance during exercise [15]. Deep learning models are adopted to identify the exerciser, and exercise metrics are defined to assess the exercise correctness.

Daily Activity Recognition Recognizing daily activities enables a variety of applications, including health monitoring, smart home, and assisted living. Various activities, e.g., walking, sitting, eating, and sleeping, can be sensed by wireless signals [4, 13, 16]. These activities can help monitor the daily activities of elderly people and patients. For example, how long people sleep, how many times people get up during sleep, when do people eat, how long they sit, how fast they walk, and whether they fall down, are essential indicators to reflect people's mental and physical states and monitor whether they have any abnormal behavior. Many proof-of-concept systems have been developed, and they use wireless signals to monitor daily activities. For instance, researchers have developed a sleeping posture recognition system by making a RFID tag matrix on the bed [17]. Most activity recognition problem can be essentially modeled as the classification problem. Therefore, the common practice is to extract features from the wireless signal indicators and train a classification model for activity recognition.

1.1.2 Applications in Security and Surveillance

Security and surveillance are critical for ensuring public and personal safety. In terms of public security, observing the presence of suspicious persons is required for protecting assets, and monitoring the crowd flow is important during big events. For personal security, illegal access to personal places and cyberattacks into digital devices can severely harm people's benefit. Wireless sensing has been applied to guarantee security in both public and personal scenarios.

Presence Detection and Crowd Counting Human detection involves detecting or counting the human-beings in a certain area. It serves many security, safety, and surveillance applications. For example, detecting whether a person enters or gets stuck in an unsafe area is important to guarantee people's safety and save people's lives in time. Crowding counting is essential for making timely and precise arrangements of crowd control in public places. In addition, many marketing research and analysis also need accurate counting of customers. Traditional human detection is mainly realized using cameras. However, cameras can intrude on people's privacy, and it may even fail to work properly due to occlusion. Therefore, many researchers have investigated the use of WiFi signals for human presence detection and crowd counting [18–21], and they can detect human presence in a non-intrusive way. Moreover, people can still be detected even when they are sheltered by the wall. The basic idea of WiFi-based human presence detection is that when a person moves into an area, the movements will cause an observable change of the signal indicator, e.g., signal amplitude, which can be used to indicate human presence. With the increasing number of people, more changes may be captured by the signal, from which we can estimate the number of present people.

Human Identification Human identification involves associating a person with a predefined identity. Various applications require access control to protect user privacy and system security. Most existing works use biometric features for human identification, requiring people to touch some sensors directly. This approach is less convenient compared with solutions that can identify a person passively. Hence, wireless signals are employed to sense the behavioral characteristics of people in a contact-free manner. Different kinds of behavioral characteristics, e.g., gait, keyboard typing, and speaking, are sensed using acoustic, RFID, WiFi, and mmWave radar signals for human identification [22–24]. Taking gait-based human identification as an example, a person's unique gait features, such as step length and step velocity, can be reflected from the signal propagation indicators. After extracting the behavioral features, an identification model is usually built using machine learning algorithms. For example, researchers composite an RFID tag array on the chest badge. Users wear the badge to measure their unique vital sign movements for human authentication [25].

Suspicious Object Detection Suspicious objects, such as explosive liquids, guns, and knives, are dangerous for people's safety, so they are banned in many public areas. Many food and liquid can be adulterated, which is hard to detect. As wireless signals can be affected by object materials and components, they are employed to detect suspicious objects and identify what kind of object it is, e.g., detecting whether a metal object appears or identifying the liquid type [26–28]. The underlying principles for object detection and identification are threefold. First, objects with different materials have distinctive reflection effects on wireless signals. For instance, metals will reflect a larger amount of RF signals than plastic materials. Second, wireless signals can penetrate objects. During the penetration, wireless signals have different reactions to different objects. For example, the traveling speed of the acoustic signal varies in different media, and the signal's energy can also be absorbed differently. Third, in addition to identifying different objects, wireless signals can also differentiate seemingly similar objects with different amounts of content, such as identifying the amount of liquid in a container. This is because different amounts of content have different impacts on the wireless signal.

1.1.3 Human-Computer Interaction

Body languages, gestures, and voice commands are basic ways for human-computer interaction. Apart from cameras, speakers, and microphones for capturing people's instructions, wireless signals can be used for human-computer interaction.

Gesture Recognition and Tracking Many studies have utilized wireless signals to achieve gesture recognition and tracking for human-computer interaction. Various kinds of gestures, including arm, leg, head, hand, finger, and eye blinking, can be recognized using wireless signals. Gesture recognition is usually conducted by defining a set of gestures, collecting the wireless signal when performing each gesture, extracting features from the signal, and training a classification model to recognize the gestures [29]. However, this approach can only identify a fixed set of gestures. To make it more scalable to other kinds of gestures, some researchers try to track the gesture with centimeter-level accuracy so that it can recover the trace of the gesture. These approaches localize the body parts when performing the gesture by measuring the time-of-flight and angle-of-arrival of the wireless signal reflected by the body part.

Speech Recognition When people talk, the mouth and surrounding facial muscles move distinctively when speaking different words that combine different vowels and consonants. Wireless signals are used to recognize what people say. In the wireless sensing literature, WiFi, RFID, and acoustic signals are employed for speech recognition. The WiFi signal's CSI amplitude is used to construct a mouth motion profile, which is then input to a machine learning model for recognizing different vowels and consonants [30]. A study has also leveraged the RFID signal to eavesdrop the speech from loudspeakers by capturing its sub-millimeter level vibration [31]. This aspect shows that RFID tags can capture sound vibration, which is a potential for speech recognition. Apart from directly recording speech by using the microphone, the acoustic signal is also used to capture the mouth and tongue movements during speaking to help improve the accuracy of speech recognition in noisy environments [32].

1.1.4 Indoor Navigation

The navigation and localization of human-beings and objects are increasingly demanded by many indoor places, such as, shopping malls, airports, and warehouses. Wireless sensing is the primary technology for indoor navigation. We have seen many commercialized systems for wireless-based indoor localization.

Human Localization Outdoor human localization has been successfully realized by GPS technology. However, GPS cannot provide accurate localization results in indoor environments. Therefore, wireless signals have been broadly investigated for indoor human localization. In the beginning, with the popularity of smartphones and their closeness with people, the sensors embedded in most smartphones, e.g., WiFi modules, speakers, and microphones, were used for localization. Later, researchers have proposed device-free localization, which does not require people to wear any sensor on the body. Acoustic, WiFi, FMCW radar, and light signals have been used for human localization applications [33–36], and the localization accuracy has been promoted from the meter level to the centimeter level. The location of a person can be determined in two ways. One approach is the model-based method, which extracts the distance or angle between the wireless sensor and target person. In Sect. 4.2, we will introduce how to extract the distance and angle information based on the signal propagation model. Another approach is the fingerprint-based method, which collects a fingerprint database for each location in advance and then maps the measured fingerprint with those in the database to obtain the location. Unlike model-based localization methods, fingerprint-based methods incur extra labor for data collection and may be ineffective under changing environments. By contrast, model-based methods are more generic and robust to environmental changes.

Object Localization In logistics applications, the precise tracking of goods in different phases of the supply chain is of vital importance for well-organized goods management, which is fulfilled by object localization and tracking [37, 38]. The RFID technology has been adopted for object tracking for a long time. The RFID tag bridges the physical object and cyberspace. Thus, the location of the RFID tag can represent the location of the object. With the development of the UHF RFID technique, multiple RFID tags can be interrogated remotely and consecutively, which expedites the process of tag interrogation. UHF RFID signals can be employed for indoor object localization, including, localizing goods stored in a warehouse. Many companies and shops, such as Uniqlo and Zara, have already embedded RFID tags on their clothes. The RFID signal phase, which can reflect the distance between the RFID reader and the tag, is mainly used for object localization and tracking [39–42].

1.1.5 Industrial Automation

In future manufacturing, we expect to achieve automated monitoring of the machines, equipment, and the environment in industrial scenarios. Wireless sensing can play a vital role towards industrial automation.

Machine and Structure Monitoring Many objects can have in-place changes, e.g., machine vibrations, structural vibration, and liquid leakage, which should be properly monitored. Abnormal vibrations indicate the presence of machine or structure damage. Liquid leakage is common in many industrial scenarios, such as liquid purification and water cooling. Meanwhile, in liquid consumption scenarios, the liquid level needs to be monitored to ensure a timely supply. Many researchers have achieved vibration measurement, liquid leakage detection, and liquid level estimation by using wireless signals, especially with the RFID signal [43, 44]. RFID tags attached to an object can vibrate along with the object. Then, the RFID signal phase changes periodically, and the vibration period can be estimated. RFID tags can also be attached to the liquid container to detect liquid leakage. In liquid leakage detection, when the liquid reaches the tag, it affects the tag's circuit impedance. By observing the signal indicators caused by the impedance change, the liquid leakage can be detected. Apart from RFID signals, radar signals can be employed for liquid level estimation owing to the millimeter-level wavelength of the radar signal, which enables a precise measurement of the distance to the liquid level.

3D Reconstruction Reconstructing 3D images of the surrounding environment is required by many applications. Robot navigation and searching tasks require the capture of the surrounding layout and obstacles to plan routes. Autonomous vehicles also need to sense nearby pedestrians, lanes, and other vehicles to avoid accidents. Lidar technology, in which laser beams are emitted to project the surrounding objects, is widely adopted in 3D scanning. However, Lidar suffers from poor weather conditions, such as rain and foggy weather. Recently, UWB and mmWave radars are more preferred for 3D reconstruction owing to its robustness to varying lighting and weather conditions [45, 46]. There have been many worldwide startups and investigations on UWB-based and mmWave-based vehicular technologies.

Temperature and Humidity Sensing Temperature and humidity sensors are usually separately used and not embedded into pervasive devices like smartphones. Therefore, researchers have proposed to leverage ubiquitous wireless devices for temperature and humidity sensing. For example, by emitting and receiving acoustic signal using the speaker and microphone of a smartphone, the sound traveling speed can be measured and correlated to temperature [47]. There are some scenarios in which traditional humidity sensors have become inconvenient to use, e.g., diaper wetness. It is quite interesting to use RFID tag for measuring diaper wetness [48] and temperature [49].

1.2 Wireless Sensing Framework

In this section, we introduce a five-layer wireless sensing framework, as shown in Fig. 1.1, to systematically understand the structure of wireless sensing. The framework demonstrates the key components in a wireless sensing system for a specific application.

The bottom layer consists of wireless signals and devices. Many devices, such as smartphones, RFID readers, WiFi routers, the universal software radio

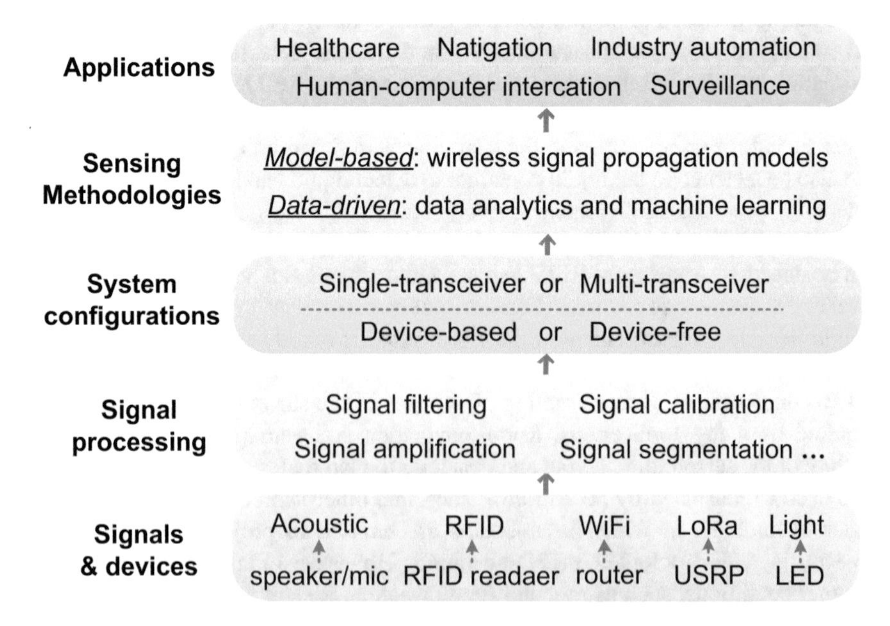

Fig. 1.1 Wireless sensing framework

peripheral (USRP), and light-emitting diode (LED), can transmit and receive wireless signals, e.g., acoustic, RFID, WiFi, and light signals. As the collected signals may involve noises and suffer from interferences, on the second layer, signal processing methods are utilized to denoise the signal and remove irrelevant disturbances. Given the selected wireless device and signal and the concerned sensing task, we need to consider the system configurations about where to deploy the wireless device and how many devices are required. The system configuration layer is the intermediate layer that connects the bottom signal-related layers with the upper task-related layers. The sensing task poses requirements and constraints on the system deployment. Therefore, we need to integrate the deployment issues when determining the proper wireless devices and signals for this sensing task. After settling down the system design and acquiring the processed signals, we apply model-based or data-driven methodologies to extract the information of the sensing target on the fourth layer. Model-based methodologies aim to derive information for the sensing target based on the wireless signal propagation models and physical laws. In data-driven methodologies, machine learning and data mining algorithms are adopted to discover underlying patterns hidden in the signal from the sensing target. The topmost layer consists of typical applications served by wireless sensing. For example, the sensed information can be employed to realize various applications, such as vital sign monitoring for healthcare, human localization for indoor navigation, gesture recognition for human-computer interaction, crowd counting for surveillance, and machine health monitoring for industrial automation.

The ultimate goal of wireless sensing is to provide the required sensing service for a particular application. Hence, we think that the design of a wireless sensing system should be application-driven. Essentially, the application requirement decides the design principle of a wireless sensing system. The first step is to figure out the application requirements, such as the sensing target and scenario, the information to sense from the target, the expected performance, and the affordable cost. Based on these requirements, the next step is to decide the type of wireless signal and the corresponding device based on these requirements. When choosing the wireless signal and device, we need to consider the configuration of the wireless device. The device can be either carried by the sensing target (i.e., device-based) or placed remotely from the target (i.e., device-free). We need to consider the feasibility of deploying the selected wireless devices under the specified application scenario. In addition, the number of wireless devices and the number of signal transceivers on the device should also be decided in terms of the requirement of system deployment cost. The final step is to apply and propose corresponding signal processing methods and sensing methodologies to obtain the sensing target's information. Meanwhile, the development of a wireless sensing system should involve hardware-level, signal-level, algorithm-level, and software-level efforts for real system implementation and use. Therefore, wireless sensing requires knowledge about the internal design of wireless devices, properties of wireless signals, principles of signal processing and data analytics methods, and the domain knowledge of a specific application. In other words, designing the wireless sensing system is a multi-disciplinary task.

1.3 Organization of the Book

The book is organized following the framework in Fig. 1.1. In Chap. 2, we will introduce the bottom two layers, including preliminary of wireless signals, common types of wireless signals, and typical signal processing methods for wireless signals. In Chap. 3, we will discuss the workflow and considerations for wireless sensing system design and present different kinds of system configurations. In Chap. 4, we will introduce the model-based and data-driven methodologies to obtain the sensing target's information. Next, we will showcase our previous studies and practices for different wireless applications in Chap. 5. Finally, we will summarize existing studies and discuss future development of wireless sensing in Chap. 6.

References

1. Liu J et al (2019) Wireless sensing for human activity: A survey. IEEE Commun Surv Tutorials 22(3):1629–1645
2. Yang B et al (2020) Acoustic-based sensing and applications: A survey. Comput Netw 181:107447
3. Pathak PH et al (2015) Visible light communication, networking, and sensing: A survey, potential and challenges. IEEE Commun Surv Tutorials 17(4):2047–2077
4. Wang H et al (2016) Human respiration detection with commodity wifi devices: do user location and body orientation matter? In: Proceedings of the 2016 ACM international joint conference on pervasive and ubiquitous computing, pp 25–36
5. Zhang F et al (2021) Unlocking the beamforming potential of LoRa for long-range multi-target respiration sensing. Proc ACM Interactive Mob Wearable Ubiquitous Technol 5(2):1–25
6. Zhang F et al (2018) From fresnel diffraction model to fine-grained human respiration sensing with commodity wi-fi devices. Proc ACM Interactive Mob Wearable Ubiquitous Technol 2(1):1–23
7. Wang C et al (2018) Rf-ecg: Heart rate variability assessment based on cots rfid tag array. Proc ACM Interactive Mob Wearable Ubiquitous Technol 2(2):1–26
8. Zhao R et al (2018) CRH: A contactless respiration and heartbeat monitoring system with COTS RFID tags. In: 15th Annual IEEE international conference on sensing, communication, and networking, pp 1–9
9. Zhao H et al (2018) Non-contact beat-to-beat blood pressure measurement using continuous wave Doppler radar. In: IEEE/MTT-S international microwave symposium-IMS, pp 1413–1415
10. Ohata T et al (2019) Non-contact blood pressure measurement scheme using doppler radar. In: 41st Annual international conference of the IEEE engineering in medicine and biology society, pp 778–781
11. Zhang F et al (2020) Exploring lora for long-range through-wall sensing. Proc ACM Interactive Mob Wearable Ubiquitous Technol 4(2):1–27
12. Zhang L et al (2018) Wi-Run: Multi-runner step estimation using commodity Wi-Fi. In: 15th Annual IEEE international conference on sensing, communication, and networking, pp 1–9
13. Han D et al (2015) Femo: A platform for free-weight exercise monitoring with rfids. In: Proceedings of the 13th ACM conference on embedded networked sensor systems, pp 141–154
14. Yang Y et al (2019) ER-rhythm: Coupling exercise and respiration rhythm using lightweight COTS RFID. Proceedings of the ACM on Interactive, Mobile, Wearable and Ubiquitous Technologies 3(4):1–24

15. Guo X et al (2018) Device-free personalized fitness assistant using WiFi. Proc ACM Interactive Mob Wearable Ubiquitous Technol 2(4):1–23
16. Wang W et al (2015) Understanding and modeling of wifi signal based human activity recognition. In: Proceedings of the 21st annual international conference on mobile computing and networking, pp 65–76
17. Liu J et al (2019) TagSheet: Sleeping posture recognition with an unobtrusive passive tag matrix. In: IEEE conference on computer communications, pp 874–882
18. Wu C et al (2015) Non-invasive detection of moving and stationary human with WiFi. IEEE J Sel Areas Commun 33(11):2329–2342
19. Li S et al (2020) WiBorder: Precise Wi-Fi based boundary sensing via through-wall discrimination. Proc ACM Interactive Mob Wearable Ubiquitous Technol 4(3):1–30
20. Xi W et al (2014) Electronic frog eye: Counting crowd using WiFi. In: IEEE Conference on Computer Communications
21. Li X et al (2020) Quick and accurate false data detection in mobile crowd sensing. IEEE/ACM Trans Netw 28(3):1339–1352
22. Hong F et al (2016) WFID: Passive device-free human identification using WiFi signal. In: Proceedings of the 13th international conference on mobile and ubiquitous systems: computing, networking and services
23. Li J et al (2020) RF-Rhythm: secure and usable two-factor RFID authentication. In: IEEE conference on computer communications, pp 361–369
24. Zhou B et al (2018) EchoPrint: Two-factor authentication using acoustics and vision on smartphones. In Proceedings of the 24th annual international conference on mobile computing and networking, pp 321–336
25. Ning J et al (2021) RF-Badge: vital sign-based authentication via RFID tag array on badges. IEEE Trans Mob Comput, Early Access
26. Ha U et al (2018) Learning food quality and safety from wireless stickers. In: Proceedings of the 17th ACM workshop on hot topics in networks, pp 106–112
27. Dhekne A et al (2018) Liquid: A wireless liquid identifier. In: Proceedings of the 16th annual international conference on mobile systems, applications, and services, pp 442–454
28. Wang C et al (2018) Towards in-baggage suspicious object detection using commodity WiFi. In: IEEE conference on communications and network security, pp 1–9
29. Wang Y et al (2020) Push the limit of acoustic gesture recognition. IEEE Trans Mob Comput 21(5):1798–1811
30. Wang G et al (2016) We can hear you with Wi-Fi! IEEE Trans Mob Comput 15(11):2907–2920
31. Wang C et al (2021) Thru-the-wall eavesdropping on loudspeakers via RFID by capturing sub-mm level vibration. Proc ACM Interactive Mob Wearable Ubiquitous Technol 5(4):article 182
32. Sun K, Zhang X (2021) UltraSE: single-channel speech enhancement using ultrasound. In: Proceedings of the 27th annual international conference on mobile computing and networking, pp 160–173
33. Yang Z et al (2013) From RSSI to CSI: Indoor localization via channel response. ACM Comput Surv (CSUR) 46(2):1–32
34. Liu K et al (2013) Guoguo: Enabling fine-grained indoor localization via smartphone. In: Proceeding of the 11th annual international conference on mobile systems, applications, and services, pp 235–248
35. Adib F et al (2014) 3D tracking via body radio reflections. In Proceeding of the 11th USENIX symposium on networked systems design and implementation, pp 317–329
36. Kuo YS et al (2014) Luxapose: Indoor positioning with mobile phones and visible light. In: Proceedings of the 20th annual international conference on mobile computing and networking, pp 447–458
37. Xie K et al (2017) Low cost and high accuracy data gathering in WSNs with matrix completion. IEEE Trans Mob Comput 17(7):1595–1608
38. Xie K et al (2016) Recover corrupted data in sensor networks: A matrix completion solution. IEEE Trans Mob Comput 16(5):1434–1448

39. Yang L et al (2014) Tagoram: Real-time tracking of mobile RFID tags to high precision using COTS devices. In: Proceedings of the 20th annual international conference on mobile computing and networking, pp 237–248
40. Zhang Y et al (2017) 3-dimensional localization via RFID tag array. In: IEEE 14th international conference on mobile ad hoc and sensor systems (MASS), pp 353–361
41. Chen X et al (2020) Eingerprint: Robust energy-related fingerprinting for passive RFID tags. In: 17th USENIX symposium on networked systems design and implementation, pp 1101–1113
42. Zhang J et al (2021) SILoc: A speed inconsistency-immune approach to mobile RFID robot localization. In: IEEE conference on computer communications, pp 1–10
43. Yang L et al (2017) Tagbeat: Sensing mechanical vibration period with cots rfid systems. IEEE/ACM Trans Netw 25(6), 3823–3835 (2017)
44. Guo J et al (2019) Twinleak: Rfid-based liquid leakage detection in industrial environments. In: IEEE conference on computer communications, pp 883–891
45. Qian K et al (2020) 3D point cloud generation with millimeter-wave radar. Proc ACM Interactive Mob Wearable Ubiquitous Technol 4(4):1
46. Chen W et al (2021) Constructing floor plan through smoke using ultra wideband radar. Proc ACM Interactive Mob Wearable Ubiquitous Technol 5(4):1–29
47. Cai C et al (2020) AcuTe: acoustic thermometer empowered by a single smartphone. In: Proceedings of the 18th conference on embedded networked sensor systems, pp 28–41
48. Sen P et al (2019) Low-cost diaper wetness detection using hydrogel-based RFID tags. IEEE Sensors J 20(6):3293–3302
49. Chen X et al (2021) Thermotag: item-level temperature sensing with a passive RFID tag. In: Proceedings of the 19th annual international conference on mobile systems, applications, and services, pp 163–174

Chapter 2
Wireless Signals and Signal Processing

2.1 Preliminaries of Wireless Signals

Wireless signals primarily used for communication purposes can be generally categorized into longitudinal waves and electromagnetic waves. In Fig. 2.1, we show the category of longitudinal waves that mainly involve infrasound, human hearing sound, ultrasound, and electromagnetic waves including radio frequency signals and light signals.

Understanding the wireless communication mechanism in terms of how wireless signals are generated and received is the first step for entering the field of wireless sensing. A typical wireless communication system is depicted in Fig. 2.2 [1]. On the transmitter (Tx) side, the information source is usually an analog source which is first fed into the analog-to-digital converter (ADC). Then, the converted digital source is coded via the source and channel coder to reduce redundancy in the source. Next, the baseband modulator assigns data bits on the baseband signal as a complex transmit symbol. The baseband consists of a range of frequencies occupied by a signal that has not been modulated to higher frequencies. Then, the digital-to-analog converter (DAC) produces analog voltages representing the real and imaginary parts of the transmit symbols. Finally, the up-converter transforms the baseband signal into the passband signal with a higher frequency by using the local oscillator.

When the receiver antenna picks up the transmitted signal, the signal is first amplified by the amplifier. Then, the passband signal is down-converted to the baseband signal by using the local oscillator on the receiver side. The baseband signal is a complex analog signal. The undesired frequency band after down-conversion is filtered out by the baseband filter. Next, the analog signal is converted to the digital signal using an ADC. Then, the baseband demodulator acquires the gross data and passes them to the decoder to obtain the original information.

In communication systems, the coder and decoder are necessary and vital elements because the information needs to be represented into 0 and 1, which is then modulated and demodulated using various methods, e.g., amplitude modulation

J. Cao, Y. Yang, *Wireless Sensing*, Wireless Networks,
https://doi.org/10.1007/978-3-031-08345-7_2

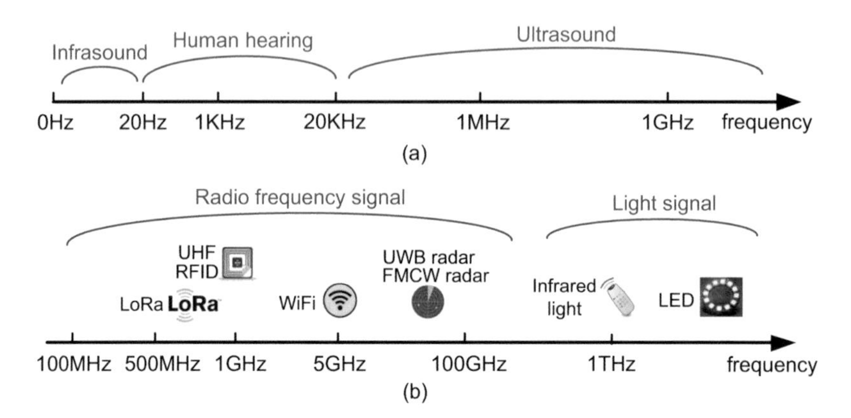

Fig. 2.1 Wireless signals in (**a**) longitudinal waves (acoustic signals with different frequencies) and (**b**) electromagnetic waves (radio frequency signals and light signals with different frequencies)

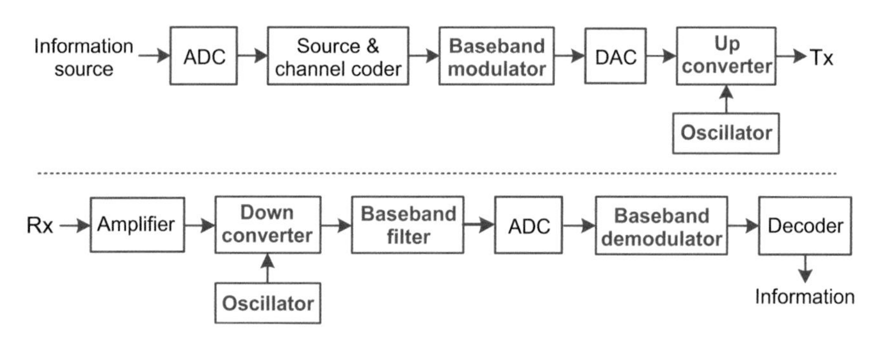

Fig. 2.2 Wireless communication diagram: the upper part involves the processing flow in the transmitter (Tx) and the bottom part involves the procedures in the receiver (Rx)

(AM) and frequency modulation (FM). As for the sensing tasks, we focus more on the formulation of baseband and passband signals, as highlighted in Fig. 2.2.

Therefore, let us introduce some basic concepts of baseband and passband signals. First, the baseband signal is a series of complex numbers. Its frequencies are mainly located around the zero frequency. The range of frequencies in the baseband signal is the bandwidth. The passband signal has the same bandwidth as the baseband signal, except that only the whole frequency band is shifted to the higher frequencies. In Fig. 2.3, we illustrate a concrete example of the up-conversion and down-conversion between the baseband and passband signals. The baseband signal $s(t)$ is converted to the passband signal $x(t)$ via

$$x(t) = s(t) \cdot cos(2\pi f_c t), \tag{2.1}$$

where f_c is the up-converted frequency which is usually much higher than the frequencies in the baseband signal. Then, the signal arrives at the receiver after a time delay of τ and becomes $y(t)$, which can be expressed as

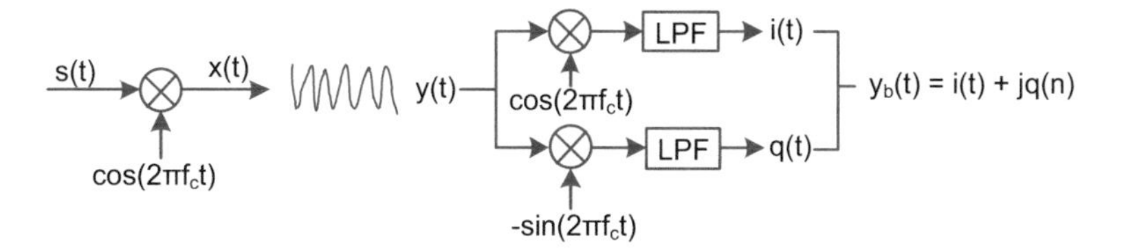

Fig. 2.3 Up-conversion and down-conversion between baseband ($s(t)$ and $y_c(t)$) and passband signals ($x(t)$ and $y(t)$), LPF refers to low-pass filter

$$y(t) = A \cdot x(t - \tau) = A \cdot s(t) \cdot cos[2\pi f_c(t - \tau)], \tag{2.2}$$

where A refers to the attenuated amplitude of the signal after transmission. By multiplying $y(t)$ with $cos(2\pi f_c t)$ and $-sin(2\pi f_c t)$, we can obtain the following result based on the product-to-sum formula:

$$y(t) \cdot cos(2\pi f_c t) = A \cdot s(t) \cdot [cos(-2\pi f_c \tau) + cos(4\pi f_c t - 2\pi f_c \tau)], \tag{2.3}$$

$$y(t) \cdot -sin(2\pi f_c t) = A \cdot s(t) \cdot [sin(2\pi f_c \tau) + sin(4\pi f_c t - 2\pi f_c \tau)]. \tag{2.4}$$

After the low-pass filter (LPF), the high-frequency parts $cos(4\pi f_c t - 2\pi f_c \tau)$ and $sin(4\pi f_c t - 2\pi f_c \tau)$ are filtered. The left part $A \cdot s(t) \cdot cos(-2\pi f_c \tau)$ and $A \cdot s(t) \cdot sin(2\pi f_c \tau)$ are employed as the in-phase (I) and quadrature (Q) components of the received baseband signal $y_b(t)$, respectively. This example shows the conversion between baseband and passband signals for a single frequency. In many communication systems, multiple subcarriers are used to send signal to simultaneously increase the throughput, such as, the orthogonal frequency-division multiplexing (OFDM) in WiFi systems.

Let us represent the received complex baseband signal $y_b(t)$ in the following equivalent form:

$$y_b(t) = A \cdot s(t) \cdot e^{-j(2\pi f_c \tau)}. \tag{2.5}$$

It shows that the amplitude A, frequency f_c, and signal propagation time delay τ jointly determine $y_b(t)$. The signal amplitude can reflect how the signal is attenuated in the wireless channel. The signal power is the squared amplitude, and it is usually represented as the received signal strength (RSS) as follows:

$$RSS = 10log_2(||A||^2). \tag{2.6}$$

The signal frequency decides the signal wavelength λ

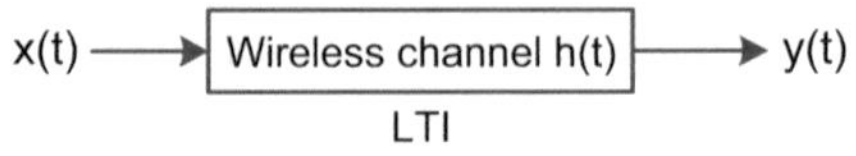

x(t) → Wireless channel h(t) → y(t)
LTI

Fig. 2.4 Wireless channel as a LTI system: $x(t)$ is the input signal, $h(t)$ is the channel impulse response, and $y(t)$ is the system output signal

$$f_c = \frac{v}{\lambda}, \tag{2.7}$$

where v is the signal traveling speed. For the complex signal $y_b(t)$, its phase is $\varphi = 2\pi f_c \tau$. For the same time delay τ, a higher f_c leads to a larger signal phase. The time delay τ is a function of the signal traveling distance $d = \tau \cdot v$. Connecting the signal phase φ and propagation distance d via the time delay τ obtains:

$$\varphi = \{2\pi f_c \cdot \frac{d}{v}\} \mod 2\pi = \{2\pi \cdot \frac{d}{\lambda}\} \mod 2\pi. \tag{2.8}$$

The signal amplitude and phase describe how the signal travels in the wireless channel. For ease of description, the channel is the medium in which the signal propagates. The channel property is characterized by its impulse response. The system's impulse response is the output when the system is input with an impulse signal. The channel impulse response $h(t)$ can be derived from the transmitted signal $x(t)$ and received signal $y(t)$. First, wireless channels are typical linear time-invariant (LTI) systems, as shown in Fig. 2.4. Therefore, we can represent the relationship between $x(t)$, $h(t)$, and $y(t)$ as follows:

$$y(t) = x(t) * h(t), \tag{2.9}$$

where $*$ refers to the convolution operation. The frequency-domain representation of Eq. (2.9) is:

$$Y(f) = X(f) \cdot H(f) \tag{2.10}$$

Then, the channel impulse response can be first retrieved in the frequency domain, i.e., $H(f) = \frac{Y(f)}{X(f)}$, and then transformed into the time domain via inverse fast Fourier transformation (IFFT).

For most acoustic and electromagnetic signals, the wireless channel is the air. The channel can also be the liquid or solid. For example, acoustic signals are mainly employed for communication in underwater environments because electromagnetic signals attenuate very fast in the liquid and solid. The propagation of wireless signals in a certain channel is generally divided into two types, namely, the line-of-sight (LOS) signal and multipath signal, as shown in Fig. 2.5. The LOS signal means that the signal travels directly from the transmitter to the receiver with one shortest path. Multipath signals involve two or more paths from the transmitter to the receiver due

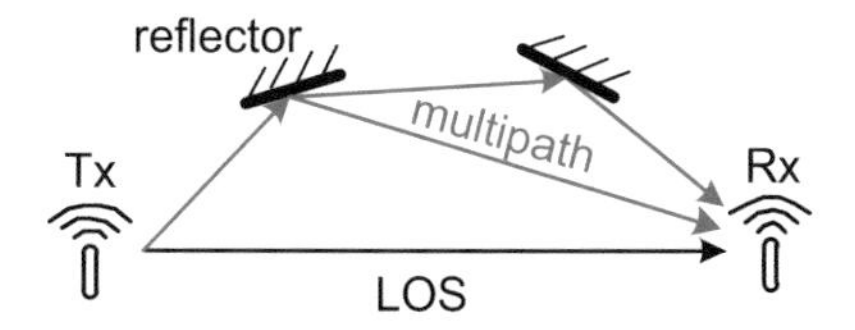

Fig. 2.5 LOS and multipath signals in the wireless channel

to the presence of reflectors. Apart from reflection, when wireless signal encounters a new medium/barrier, diffraction, scattering, and refraction can also happen.

In Fig. 2.2, the wireless signal used for communication is encoded and decoded to convey messages. However, for the sensing purpose, we may only need to send and receive the signal without the coder and decoder. Next, we will introduce three typical forms of the transmitted signal utilized in wireless sensing, including the continuous wave (CW), the pulse signal, and the FMCW signal.

The first form is the CW signal, which is a single-tone signal with a carrier frequency f.

$$x(t) = cos(2\pi f t) \tag{2.11}$$

The CW signal is widely employed in acoustic and RFID sensing systems [2, 3]. When the CW signal propagates from the transmitter to the receiver, it travels in different paths due to the presence of multiple reflectors in the environment. Meanwhile, the signal experiences a certain propagation delay τ along each path. Thus, the impulse response of the propagation channel can be expressed as

$$h(t) = \sum_{l=1}^{L} \alpha_l \cdot \delta(t - \tau_l), \tag{2.12}$$

where L is the total number of paths, and α_l is the attenuation of the signal power. Then, the received signal is expressed as

$$y(t) = \sum_{l=1}^{L} \alpha_l \cdot cos[2\pi f(t - \tau_l)]. \tag{2.13}$$

After down-converting $y(t)$, as depicted in Fig. 2.3, the baseband signal can be formulated as

$$y_b(t) = \sum_{l=1}^{L} \alpha_l \cdot e^{-j(2\pi f \tau_l)}. \tag{2.14}$$

The second form is the pulse signal, which is mostly employed in the UWB radar [4]. The pulse signal adopts a Gaussian pulse wave $s_{tx}(t)$ as follows:

$$s_{tx}(t) = e^{-\frac{(t-T_p/2)^2}{2\sigma_p^2}}, \tag{2.15}$$

where T_p is the time duration of the pulse signal, σ_p^2 is the variance corresponding to the -10 dB bandwidth.[1] The transmitted signal $x(t)$ is a series of pulse signals with a certain interval among them. Each segment of the pulse signal is called a frame. The up-converted transmitted signal $x_n(t)$ is represented as

$$x_n(t) = s_{tx}(t - nT_s) \cdot cos(2\pi f(t - nT_s)), \tag{2.16}$$

where n is the nth frame, $T_s = 1/f_p$ is the duration of the frame, i.e., the interval between two pulse signals, and f_p is the pulse repetition frequency. The pulse signal $s_{tx}(t)$ is up-converted to the carrier frequency f. In Fig. 2.6, we depict one pulse signal of $x_n(t)$. The channel impulse response of the nth frame is formed as

$$h_n(t) = \sum_{l=1}^{L} \alpha_l \cdot \delta(t - \tau_l - \tau_l^D(nT_s)), \tag{2.17}$$

where $\tau_l^D(nT_s)$ is the time delay caused by the Doppler frequency shift of the lth path. Then, the received signal of the nth frame can be expressed as

$$y_n(t) = \sum_{l=1}^{L} \alpha_l \cdot cos(2\pi f(t - nT_s - \tau_l - \tau_l^D(nT_s))) \cdot s_{tx}(t - nTs - \tau_l - \tau_l^D(nT_s)). \tag{2.18}$$

After down-conversion, the baseband signal of the nth frame is given by

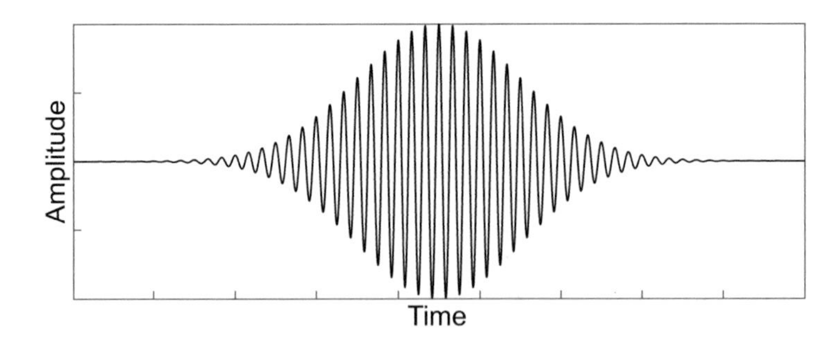

Fig. 2.6 Waveform of the Gaussian pulse signal $x_n(t)$

[1] $\sigma = 1/(2\pi B_{-10dB}(log_{10}e)^2)$.

$$y_{b_n}(t) = \sum_{l=1}^{L} \alpha_l \cdot e^{j2\pi f(\tau_l + \tau_l^D(nT_s))} \cdot s_{tx}(t - nTs - \tau_l - \tau_l^D(nT_s)). \tag{2.19}$$

In the pulse signal, the time in each pulse signal is denoted as the "fast time". The pulse signal is very short and transmitted in a repetitive way with a fixed time interval between adjacent pulses. Each pulse frame is called the "short time", which corresponds to the physical timestamp.

The third form is the FMCW signal, whose frequency linearly sweeps in each chirp of signal $s_{tx}(t)$. FMCW has been widely employed in acoustic, LoRa, and mmWave signals [5]. The sweeping frequency is expressed as $f(t) = f_c + kt$, where f_c is the starting frequency, and k is the slope. By taking the integral of $f(t)$ along time, the FMCW chirp can be represented as

$$s_{tx}(t) = e^{j2\pi(f_c t + \frac{1}{2}kt^2)}. \tag{2.20}$$

The transmitted signal is a set of consecutive FMCW chirp signals, which is given by

$$x_n(t) = \sum_{n=1}^{N} s_{tx}(t - nT_s), \tag{2.21}$$

where N is the total number of chirps, and T_s is the duration of the chirp. The waveform of one FMCW chirp is shown in Fig. 2.7. The frequency of the FMCW chirp over time is depicted in Fig. 2.8. The channel impulse response of the nth chirp is the same as illustrated in Eq. (2.17). Then, the received FMCW signal of each chirp is given by

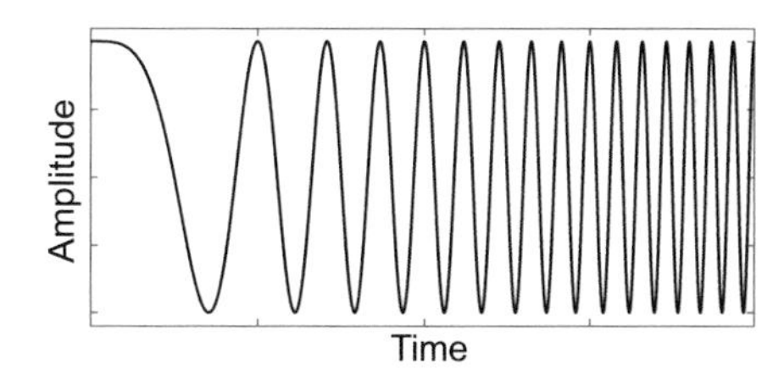

Fig. 2.7 LFM chirp in the time domain

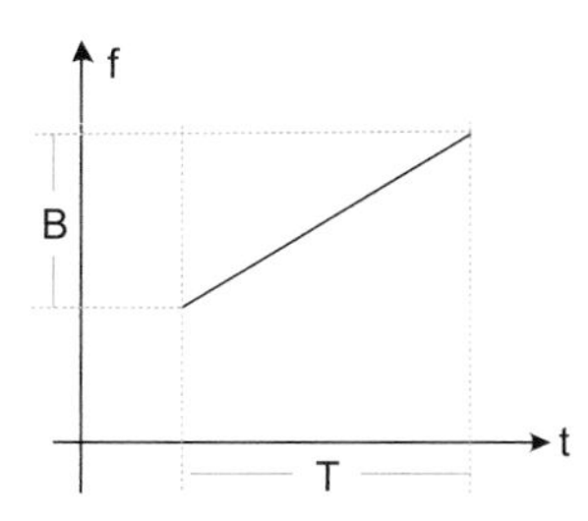

Fig. 2.8 The frequency of LFM chirp

$$y(t) = \sum_{l=1}^{L} \alpha_l \cdot e^{j2\pi[f_c(t-\tau_l)+\frac{1}{2}k(t-\tau_l)^2]}. \quad (2.22)$$

At the receiver side, a mixer is added to multiply $y(t)$ by $s_{tx}(t)$ to acquire the inter-frequency (IF) signal as follows:

$$y_{IF}(t) = \sum_{l=1}^{L} \alpha_l \cdot e^{j[2\pi(f_c\tau_l+k\tau_l t-\frac{1}{2}k\tau_l^2)]}. \quad (2.23)$$

Next, we can obtain the in-phase (I) and quadrature (Q) of the chirp signal corresponding to the real and imaginary parts of the baseband complex signal, as follows:

$$I = Re\{y_{IF}(t)\}, Q = Im\{y_{IF}(t)\}. \quad (2.24)$$

There are also other forms of $x(t)$ depending on the requirements of the wireless system. For example, we can apply the Zadoff-Chu sequence [6] to $x(t)$ to improve the signal-to-noise ratio (SNR). The aforementioned signals are transmitted upon a single carrier frequency at a single time. In practice, wireless signals also can be transmitted in multiple carrier frequencies simultaneously.

2.2 Common Types of Wireless Signals

In this section, we will introduce different kinds of wireless signals and how they are utilized in wireless sensing.

2.2.1 *Acoustic Signal*

As shown in Fig. 2.1a, we classify acoustic signals into infrasound, human hearing sound, and ultrasound signals. Ultrasound, which is defined as "sound at frequencies greater than 20 KHz" by the American National Standards Institute, is widely used in many industrial and medical applications, e.g., the Sonar for object detection and ranging and acoustic microscopy for internal body structure imaging [7]. The sounds heard by human beings are primarily used for communication and entertainment. In the past decades, with audio infrastructures embedded into the Internet of Things (IoT) and mobile devices, acoustic transceivers, i.e., speakers and microphones, have been developed beyond their primary functions, from recording and playing sounds to fulfilling sensing tasks, such as human tracking and activity recognition, object detection, and environment state estimation.

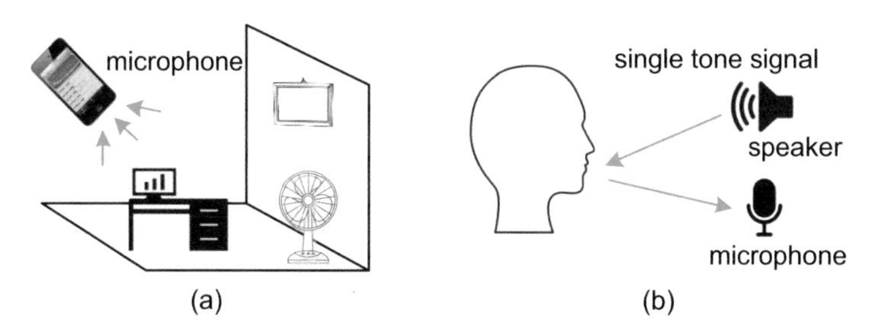

Fig. 2.9 Examples of (**a**) passive and (**b**) active acoustic sensing

Acoustic sensing can be generally divided into two means, including: passive and active sensing. In passive acoustic sensing, the acoustic receiver records the acoustic signal in the environment in a passive way. No dedicated acoustic transmitter is required to send any specified signal. As illustrated in Fig. 2.9a, the microphone records the background sound in the room, extracts a fingerprint for each room, and enables room-level localization. In active acoustic sensing, the acoustic transmitter actively sends out a designated signal to the receiver. As shown in Fig. 2.9b, the speaker actively sends out a single tone signal[2] that can be reflected by the human mouth during talking. The reflected signal is then received by the microphone and processed to analyze the talking behavior.

Compared with active acoustic sensing, passive sensing is simpler to deploy as it only requires a receiver. By contrast, active sensing needs both the acoustic transmitter and receiver and a specialized design of the transmitted signal. Nevertheless, active sensing is more promising for achieving various applications because we have the flexibility to design acoustic signals to provide the expected sensing capability, e.g., the resolution of distance measurement. Different kinds of acoustic devices have been used in active sensing, including the common mobile devices (e.g., smartphones, watches, and even the recently popular audio assistant Amazon Alexa) and specialized acoustic sensors (e.g., ultrasound acoustic modules). The transmitted signal is mainly in the frequency band of 18–22 KHz for mobile devices because the upper bound of most commercial speakers and microphones is under 22 KHz, and the sound above 18 KHz is less audible and disturbing to humans. For signals above 22 KHz with a broader bandwidth, specialized acoustic sensors are usually needed.

Acoustic signals own unique features for sensing. First, the propagation speed of acoustic signals in the air is much lower than that of electromagnetic signals. When measuring the signal propagation time for ranging tasks, acoustic signals can provide higher accuracy. For example, if there is a 0.001 s error in time measurement, the range error estimated by the acoustic signal with the speed of 340 m/s is 0.34 m, while the error becomes as large as 3×10^5 m for electromagnetic

[2] The single tone signal can be expressed as $cos(2\pi f t + \phi)$.

signals (signal propagation speed of 3×10^8 m/s). Second, the acoustic signal can well penetrate the liquid and solid channels, and different media have distinctive effects on the acoustic signal. For example, sound speed is largely affected by the properties of the medium, e.g., the material and temperature. In the air, the sound speed v is a function of temperature (T), i.e., $v^2 \approx 403\,\mathrm{T}$, where T is in the unit of Kelvin [8]. v is higher in the solid medium than that in the liquid or gas medium. Hence, acoustic signals can be used to identify objects with different components and estimate the environment temperature. Third, acoustic absorption, acoustic dispersion, and acoustic resonance phenomena are adopted for many object sensing applications. However, acoustic-based sensing suffers from a critical issue, which is the small sensing range. Existing investigations on the use of acoustic signals for sensing are mainly achieved within 1–2 m. The small sensing range requires the target to locate nearby the acoustic transceivers, which limits the application of acoustic-based sensing in large-scale scenarios.

2.2.2 *RFID Signal*

RFID signals are classified according to their working frequency bands: low frequency (120–150 KHz), high frequency (13.56 MHz), and ultra-high frequency (UHF) (865–868 MHz in Europe and 902–928 MHz in North America) [9]. As low-frequency and high-frequency signals mainly work in small ranges (0.1–1 m) based on the inductive coupling effect, the sensing range limits them to be adopted in long-range sensing scenarios. Therefore, UHF signals, which have a longer communication range, are more widely used for sensing.

As shown in Fig. 2.10, the typical RFID-based sensing system usually consists of the RFID reader, RFID antenna, and RFID tag. The RFID reader in Fig. 2.10 works in a full-duplex way, in which the reader transmits signals via the antenna to power up the tag meanwhile receiving the signal backscattered from the tag. There are also simplex and half-duplex RFID transceivers. In the simplex way, the RFID transceiver can only act as the transmitter or receiver. In the half-duplex manner, the RFID reader can transmit and receive the signal, but not in a simultaneous way. The reader can only transmit first and then switch to receive the backscattered signal from the tag. RFID tags can be active or passive. Active tags need to be charged with an external battery, whereas passive tags can harvest energy from the reader antenna. Passive tags are more popular because they are lightweight and battery-free.

By attaching RFID tags to the target, the target motions and properties can be inferred from the backscattered line-of-sight (LOS) signal from the tag. RFID

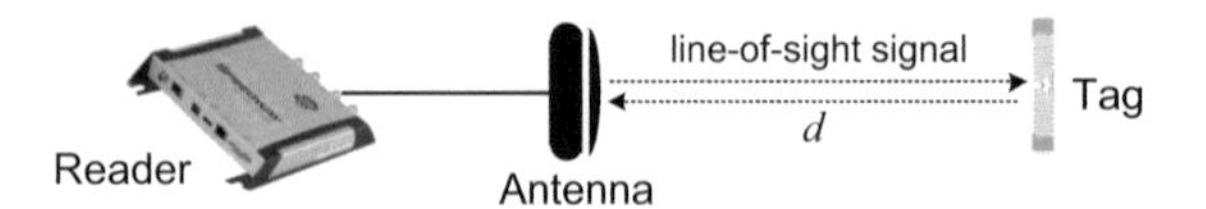

Fig. 2.10 Components of a RFID-based sensing system: RFID reader (e.g., ImpinJ), RFID antenna, and RFID tag

signals are widely employed for inventory and target tracking. In inventory management, by reading the tag ID acting as a unique identifier for each item, we can check the existence of multiple items. When there are multiple targets, each target's information can be separately extracted based on the tag ID. The low cost of the tag also adds value to large-scale sensing. Therefore, RFID-based wireless sensing has received more attention in logistics applications, which require the localization and tracking of numerous objects. To this end, RFID sensing is a key enabling technology to realize unmanned shops and automatic warehouses.

Target tracking is usually accomplished by estimating and tracking the distance or angle of arrival (AOA) between the RFID reader and tag. Both the signal amplitude and phase can be leveraged to calculate the distance. However, the signal phase is more robust for distance measurement and can also be employed to estimate the AOA of the tag.[3] The change in distance between the tag and reader can be retrieved from Eq. (2.8). As for commercial RFID readers, a phase shift ϵ is added in Eq. (2.8) in consideration of the hardware imperfection and environmental effects, i.e.,

$$\varphi = \{2\pi \cdot \frac{d}{\lambda} + \epsilon\} \mod 2\pi. \tag{2.25}$$

By deploying multiple antennas, the phase difference $\Delta\varphi$ among different antennas can be used to estimate the angle-of-arrival (AOA) of the tag signal.[4] In addition to motion sensing, the RFID tag can also recognize the property of the attached object because the object material or component can influence the impedance change on the tag circuit. Therefore, RFID tags are also employed for material or content recognition.

2.2.3 *WiFi Signal*

WiFi is a family of wireless network protocols. Most current WiFi devices follow the IEEE 802.11n standard, which supports 20/40 MHz bands at 2.4/5 GHz frequency bands within the unlicensed ISM (industrial, scientific and medical) spectrum. The recent IEEE 802.11ax standard expands the bandwidth to 160 MHz, and the IEEE 802.11ad standard is developed to provide as high as 60 GHz frequency and several gigahertz of bandwidth. In terms of increasing the throughput, current WiFi access points (APs) support the multiple-input multiple-output (MIMO) with multiple transmitting and receiving antennas. Orthogonal frequency division modulation (OFDM) is adopted to encode data on multiple carrier frequencies while ensuring

[3] In Sect. 4.2.1.1, we will explain why the signal amplitude does not perform robustly in distance estimation.

[4] AOA estimation will be discussed in detail in Sect. 4.2.1.2.

Table 2.1 Frequency and wavelength of WiFi signals

Standard	IEEE 802.11b/g/n	IEEE 802.n/ac/ax	IEEE 802.ad/aj
Frequency	2.4 GHz	5 GHz	60 GHz
Wavelength	12.5 cm	6 cm	5 mm

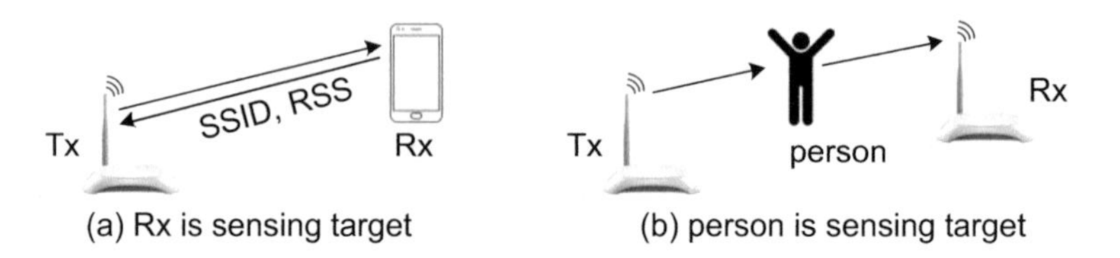

Fig. 2.11 Deployment of wireless sensing systems: (**a**) Rx can be smartphone or other mobile devices with WiFi module, which is carried by the sensing target. (**b**) The sensing target (e.g., the person in the middle) does not carry Tx or Rx

higher resilience to RF interference and low multipath distortion. In Table 2.1, we list the key WiFi standards and the corresponding signal frequencies.

WiFi sensing systems are usually deployed in two ways. In the first way, as shown in Fig. 2.11a, a WiFi AP acts as the transmitter, and the mobile device (e.g., smartphone and laptop) acts as the receiver. The receiver itself is the sensing target. Under this setting, the receiver can also be carried by the sensing target in which the receiver's state represents the target. When the transmitter and receiver communicate with each other, we can obtain the application-layer information, e.g., the service set identifier (SSID) of the receiver and the RSS of the received signal, to sense the target. In the second way, as depicted in Fig. 2.11b, the transmitter and receiver are fixed. The target affects the signal propagating from the transmitter to the receiver. This kind of deployment is called device-free sensing in which the target does not need to carry any sensor.[5]

In WiFi sensing, the RSS is first used for many sensing applications because RSS is exposed in the application layer of the WiFi standard, and it can be easily obtained. For example, indoor localization is achieved by estimating a rough distance between the target device to the WiFi AP from the RSS. However, RSS is the aggregation of all signal paths in the environment, suggesting that Eq. (2.6) can be extended to the following:

$$RSS = 10log_2(||A||^2),\ A = \sum_{i=1}^{L} ||A_i||e^{-j\varphi_i}, \tag{2.26}$$

where L is the number of signal paths. However, the target mainly affects one signal path. Thus, the RSS is not a robust indicator for estimating the target's state. To acquire fine-grained information of the target, the physical-layer channel impulse

[5] We will discuss in detail about device-free sensing in Sect. 3.2.

response $h(t)$ is used for WiFi sensing. The WiFi signal adopts the OFDM in which multiple subcarriers with different central frequencies are employed to send the WiFi signal simultaneously. To this end, the frequency-domain representation of the channel impulse response called channel state information (CSI) $h(f,t)$, as represented in Eq. (2.27), is popularly leveraged in WiFi sensing. In CSI, both the signal amplitude and phase can be used for sensing the target.

$$h(f,t) = \sum_{i=1}^{N} A_i(f,t)e^{-j\varphi_i(f,t)}. \tag{2.27}$$

The upgrade of WiFi standards steadily improves the sensing resolution of WiFi signals. As depicted in Eq. (2.7), a higher signal frequency f means a shorter signal wavelength λ. Signals with a small λ are more sensitive to the movements of tiny objects with tiny movements because signals with a large λ will bypass the object with a small size. Consequently, fewer signals will be reflected from the object. Meanwhile, signal indicators, e.g., amplitude and phase, experience more significant changes by a signal with a larger λ. To interpret the capability of WiFi signals for sensing, researchers draw from Fresnel zone theory to model how the target location and orientation can affect sensing performance [10, 11].

The key advantage of WiFi sensing lies in the wide availability and continuous upgrade of various WiFi modules. First, smartphones, laptops, and many other IoT devices all support the WiFi connection. By utilizing these devices, we can achieve various sensing tasks, such as WiFi-based crowd counting, user localization, and navigation [12]. Apart from sensing the crowd and large-scale human activities, with the increasing frequency and bandwidth of WiFi signals, we can also detect tiny body movements, e.g., finger gestures, respiration, and even the heartbeat. Meanwhile, current WiFi APs and network interfaces (NICs) are equipped with more than three antennas, which can also improve the sensing resolution and accuracy.

2.2.4 LoRa Signal

LoRa ("Long Range") is featured as a low-power and long-range communication technique. LoRa mainly works at the sub-gigahertz frequency bands, e.g., 433–434 MHz, 863–873 MHz, and 915–928 MHz in different regions. LoRa adopts the chirp spread spectrum (CSS) modulation scheme in which the transmitted data are modulated in linear frequency modulation (LFM) chirps as follows:

$$x(t) = e^{j2\pi f_c t + j\pi k t^2}, \tag{2.28}$$

where f_c refers to the center frequency. $f(t) = f_c + kt$, where $k = \frac{B}{T}$, which is the frequency sweep rate of the chirp, B is the bandwidth, and T is the time duration

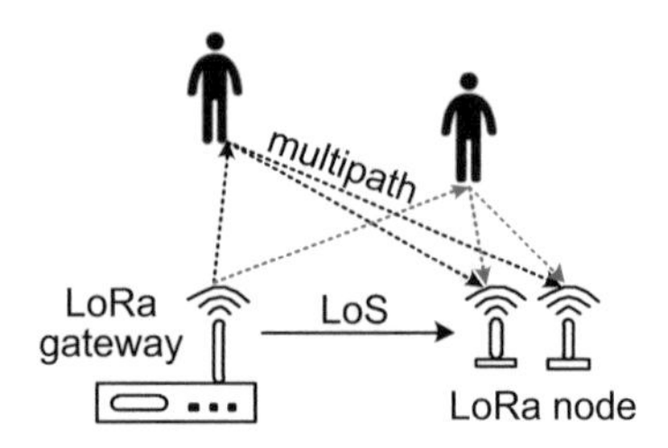

Fig. 2.12 LoRa sensing system set-up and real-deployment: LoRa gateway is the transmitter, and LoRa node is the receiver

of a chirp. The time-domain representation of the signal expressed in Eq. (2.28) is shown in Fig. 2.7.

A typical LoRa sensing system is illustrated in Fig. 2.12. The transmitter is the LoRa gateway, and the LoRa node is the receiver. The sensing targets, e.g., the two persons in Fig. 2.12, brings dynamic multipath signals to the received signal. The LoRa node is usually equipped with two antennas to cancel the carrier frequency offset and sampling frequency offset between the transmitter and receiver.

The design of LoRa signals enables the carried information to be successfully decoded even when the SNR is below −20 dB. This feature is the most promising one for LoRa sensing, namely the long sensing range. Existing works have realized LoRa-based human activity sensing in 25–50 m [13–15]. However, the long communication range of LoRa signals also brings a technical challenge. More surroundings exist in a large sensing area, and there may be multiple targets to sense. As a result, the multiple targets' signals will interfere with each other in the received signal. Many solutions have been proposed to solve this problem. We will introduce one of the most representative solutions called beamforming in Sect. 4.2.1.6.

2.2.5 *Radar Signal*

Radio detection and ranging (Radar) employs radio signals to estimate the distance, angle, or velocity of the target. There are various kinds of radar signals, such as the continuous wave (CW) radar signal, ultra-wideband (UWB) radar signal, and frequency-modulated carrier wave (FMCW) radar signal. Moreover, radar is generally divided into two categories. One of them is the CW radar that continuously transmits high-frequency signal; it measures the phase difference between the transmitted signal and received signal, which is used to measure the traveling distance of the signals, as expressed in Eq. (2.8). The frequency of the CW radar signal is constant. When the target is moving at a radial velocity, the signal reflected by the target will be shifted by a Doppler frequency. Thus, CW radars, which specifically aim to measure the target Doppler frequency, is also called Doppler radar. In the other category of radars, instead of CW, they repetitively transmit a pulse or chirp signal. As shown in Fig. 2.13, the pulse is emitted in a fixed interval, which is called the pulse repetition period. In the silent period between two pulses, echoes reflected from targets are received. We can calculate the distance between the

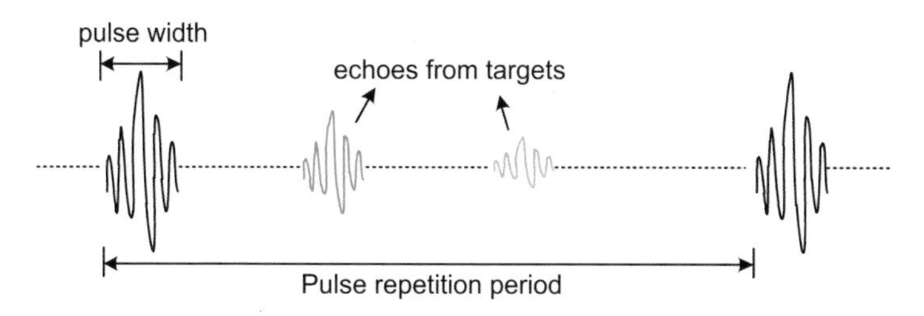

Fig. 2.13 Mechanism of the pulsed radar

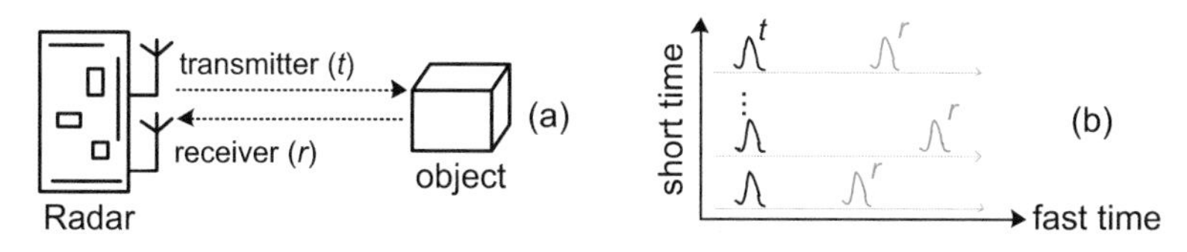

Fig. 2.14 Radar-based sensing (**a**) monostatic system setup, (**b**) signal dimensions of IR-UWB radar in fast time and short time

target and radar transmitter by deriving the time interval between the transmitted and received signals.

Most radars follow the monostatic deployment in which the radar transmitter and receiver (also called transceiver) are implemented on a single circuit, as shown in Fig. 2.14. The transmitted signal propagates to an object, is reflected by the object, and propagates back to the receiver. As such, the propagation path of the pulse signal is twice the distance between the radar transceiver and the object. For pulsed radars, the dimension along the pulse signal is called the fast time. Fast time reflects how long the pulse signal travels from the transmitter and propagates back to the receiver. The accumulation of pulse signal is denoted as short time, which is the physical time, as shown in Fig. 2.14b. The pulse signal can adopt different forms. For the UWB radar, the pulse signal has a Gaussian distribution. For the FMCW radar, the frequency of the pulse signal, which is also called the chirp signal, sweeps linearly over time.

Radar sensing systems have many prevailing features. First, given the high signal frequency, the wavelength of the radar signals is quite small, which is around the millimeter (mm) level. As such, radars whose signal frequency is above around 20 GHz, are often called mmWave radars. Consequently, radar signals can detect small objects with tiny movements in high sensitivity. Second, many radars have a relatively larger bandwidth, which is usually around 2–4 GHz, than the RFID, WiFi, and LoRa signals. As the ranging resolution Δd of signal is determined by its bandwidth, i.e., $\Delta d = \frac{c}{2B}$, where c is the light speed, and B is the bandwidth. Radar signals can achieve fine-grained ranging resolutions. Third, radar antennas can be quite small, which enables the multi-antenna design on a single radar circuit board. With multiple antennas, we can measure the angle of the target. Moreover, with the recent development of 5G technology, mmWave frequency bands in the 24–29 GHz are globally opened up and licensed with more off-the-shelf commercial and cheap mmWave units available in the market. Many operators are deploying 5G networks with the mmWave spectrum. We have witnessed the intensive study of the mmWave for automobile and industrial applications [16, 17]. It is believed that radar sensing systems will be actively investigated for new and challenging sensing tasks in the near future.

2.2.6 Light Signal

Light signals can be divided into infrared light and visible light. Infrared light has been widely used for temperature measurement, surveillance, night vision, and so on. Lidar (light detection and ranging) also employs near-infrared light and enables precise 3D scanning of the surrounding environments. However, devices that support infrared light is costly. For the sake of pervasive sensing, visible light, as it is the most common light in our daily life, also has the potential for sensing. Visible light sensing systems reuse the existing lighting infrastructure to save hardware costs.

The backbone principle of visible light sensing is to employ the shadow caused by human presence. Due to the nanometer-level wavelength of the light signal, the human body can completely block the light and result in a shadowing area. Visible light sensing is initially realized by densely installing LED lights on the ceiling and photodiodes on floors, as shown in Fig. 2.15a. LED lights transmit the light signal to photodiodes that record light intensity changes. By projecting the 3D human pose onto the 2D shadow on the floor, we can track different body gestures. However, as multiple LED lights simultaneously emit signals, composite and diluted shadows make it difficult to extract clear body gestures. Thus, we need to separate light signals sent from different LED light sources. To achieve this setup, the most common solution is to modulate the light signal with a unique frequency for each light source [18].

The design of the visible light sensing system in Fig. 2.15a has a practical limitation: it requires deploying numerous light sources and photodiodes. This setup increases the system complexity and cost. Therefore, many solutions have been

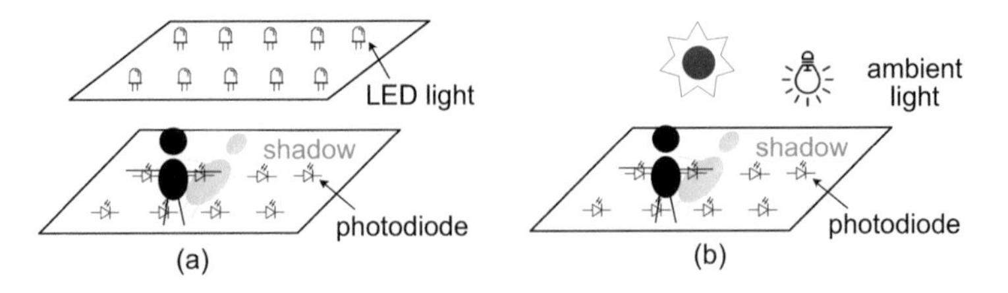

Fig. 2.15 Visible light sensing system using (**a**) LED light and photodiodes, (**b**) ambient and photodiodes

proposed to ease the limitation. For example, instead of capturing the shadow pattern from the photodiodes, a virtual shadow map is obtained from the LED panel, which can reduce the number of photodiodes on the floor. Another method is to employ the ambient light in the environment as the light source, as shown in Fig. 2.15b, so that the installation of the designated LED light sources can be saved. Visible light sensing can be further realized in a battery-free way using the solar cell [19]. Solar cells can harvest energy from various light sources. Meanwhile, the change in light condition can be reflected from the voltage change of the solar cells.

Visible light sensing entails a critical problem originating from the property of the light signal. The light signal's wavelength is extremely small, rendering the failure to penetrate barriers and shelters. If an obstacle blocks the sensing target, then the light signal cannot detect it. Hence, visible light sensing is more proper for sensing in a small local area.

2.3 Wireless Signal Processing Solutions

Many wireless sensing systems utilize prevalent and low-cost wireless devices, e.g., smartphones and WiFi routers, to make the system ubiquitous and easily accessible to people. However, as commercial off-the-shelf wireless devices are mainly designed for communication purposes, the embedded hardware components may introduce some noises in the wireless signal for sensing. Furthermore, some applications require high resolution in measuring the physical indicators, but existing wireless devices and regulations are constrained to provide demanded resources, e.g., the number of transceivers on the device and bandwidth. In this section, we will introduce the signal processing methods to deal with the key imperfections and limited resources in wireless signals.

2.3.1 Signal Processing to Remove Signal Noises

The wireless signal indicators, including signal amplitude, phase, and frequency response, are prone to many noises.

The signal amplitude is mainly affected by the amplifier in wireless devices. The amplifier, which is used to convert a low-power signal into a high-power one, is important and equipped in almost every wireless device. Amplifiers can be categorized into different kinds, namely, linear and nonlinear amplifiers, power amplifier (PA), low noise amplifier (LNA),[6] high power amplifier (HPA), programmable gain amplifier (PGA), etc. In the transmitter, the PA is employed to increase the signal

[6] LNA is widely used in RF receivers. It amplifies a very low-power signal without degrading the signal-to-noise ratio.

Table 2.2 Low-pass/smoothing filters

Filter	Overview
Butterworth filter	Has a slower roll-off[a]
Chebyshev filter	Has a shaper roll-off
Mean/Median filter	Use mean/median value in a window for smoothing
Savitzky-Golay filter	Use least-square fitting for smoothing
Wavelet filter	Use DWT[b] to decompose and get approximation parts

[a] Roll-off is the steepness of frequency response
[b] DWT: discrete wavelet transform; DWT decomposes the signal into detail and approximation parts

power level. The LNA and PGA are usually used at the receiver side. The amplifier inherently suffers from many noises, such as thermal noise and transistor noise. When amplifying the signal at the receiver side, the amplifier cannot perfectly compensate for the loss of power caused by the attenuation when the signal travels from the transmitter to the receiver. Thus, the uncertainty in the power amplifier adds noises to the signal amplitude.

There are mainly two kinds of noises: high-frequency noises and outliers. To remove high-frequency noises, low-pass or smoothing filters are commonly employed. There are many kinds of low-pass and smoothing filters, including the Butterworth/Chebyshev low-pass filter, mean/median filter, Savitzky-Golay filter, and wavelet filter. We list the basic principle of several popular filters in Table 2.2.

A straightforward method for removing amplitude outliers is to detect the signal points whose value is far from the neighboring points. Then, the outlier points can be replaced by their neighboring points' mean/median value. This idea is adopted by the Hampel filter, which is widely used to remove outliers. The Hampel filter first calculates the median (med) and standard deviation (std) of the signal in a certain window. If a point has a value larger than $med + 3 \times std$ or smaller than $med - 3 \times std$, it will be regarded as an outlier. Then, this outlier is substituted by the median value of the window.

The noises in the signal phase are mainly caused by the imperfect local oscillator in the wireless transmitter/receiver. First, during local oscillator quadrature conversion, IQ imbalance is introduced due to two reasons [20]: (1) A phase mismatch exits between the I and Q components of the local oscillator signal, which means that the I and Q components are not exactly 90°. (2) The gain of the I and Q components can be different, which leads to amplitude mismatch. We represent the IQ imbalance in Fig. 2.16, where ϵ_r and $\Delta\varphi_r$ refer to the local oscillator gain and phase errors, respectively. IQ imbalance causes a nonlinear phase error on signals at different subcarriers [21].

Apart from IQ imbalance, for bistatic wireless systems in which the transmitter and receiver are separately placed, the local oscillators in the transmitter and receiver can be unsynchronized, resulting in several issues: (1) an initial phase difference will be introduced as the phase-locked loop offset (PPO). (2) There would be a time shift after sampling the signals in ADC, subsequently causing the sampling frequency offset (SFO). Although some receivers have the SFO corrector attempting

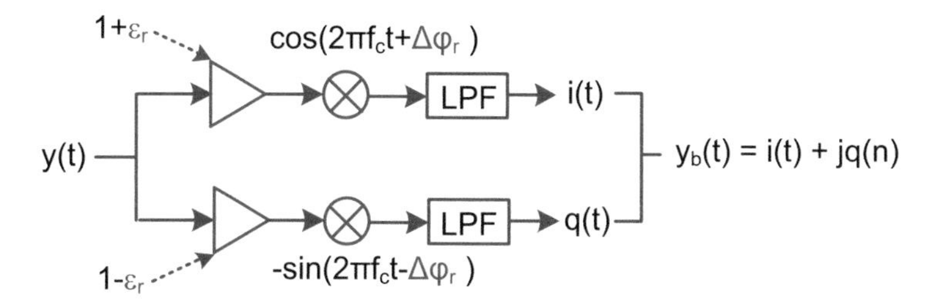

Fig. 2.16 IQ imbalance due to gain (ϵ_r) and phase ($\Delta\varphi_r$) errors in the receiver

to calibrate the SFO, there remains a residual SFO. (3) For wireless communication systems, there exists a packet detection module that can detect the arrival of signal. However, unsynchronized oscillators also cause a packet detection delay (PDD). (4) The central frequencies of the transmitter and receiver cannot be precisely synchronized. Then, a central frequency offset (CFO) is introduced. In summary, if all of the abovementioned noises are involved, the measured phase φ_m can be expressed as follows:

$$\varphi_m = \varphi_r + \psi + 2\pi(f \cdot \sigma + \Delta f) + Z, \tag{2.29}$$

where φ_r is the real phase, ψ refers to the error from the IQ imbalance, σ is the timing offset due to SFO and PDD, and Δf denotes the CFO. All of the sources of errors lead to highly distorted phase measurements.

Signal phase is a basic piece of information for measuring the time delay of signals traveling from the transmitter to the receivers. If the phase values are wrongly measured, then the latter estimation of the angle of arrival (AOA), angle of departure (AOD), and time of flight (TOF) using model-based methodologies will be biased. This issue will not only affect the realization of the localization applications that require the absolute position of the target, but also those applications that need to measure the relative movement changes, e.g., gesture tracking. Hence, phase noises should be properly eliminated.

There are two methods for removing signal phase noises. The first method is to harness multiple receiving antennas to obtain the difference of the signal phase from two antennas [22, 23]. Many wireless devices are equipped with multiple antennas, such as WiFi network interface cards and LoRa nodes. The benefit of using multiple antennas is that the IQ imbalance, CFO, and SFO are the same for all antennas on the same wireless device. If we subtract the signal phase of two antennas, then ψ, σ, and Δf are eliminated in Eq. (2.29). Specifically, for two antennas deployed with l distance between them, we express the signal phase with the AoA of θ from the first antenna as follows: $\varphi_{m1} = \varphi_{r1} + \psi + 2\pi(f \cdot \sigma + \Delta f) + Z_1$. The signal phase of the second antenna is expressed as follows: $\varphi_{m2} = \varphi_{r1} + \frac{l \cdot sin\theta}{\lambda} + \psi + 2\pi(f \cdot \sigma + \Delta f) + Z_2$. Then,

$$\varphi_{m2} - \varphi_{m1} = \varphi_{r1} + Z_2 - Z_1 + \frac{l \cdot sin\theta}{\lambda}. \tag{2.30}$$

As depicted by Eq. (2.30), the phase noises caused by IQ imbalance, CFO, and SFO are cancelled out by the phase difference. Meanwhile, the signal AOA information is added. The phase difference method is a simple yet effective way for removing phase noises.

However, we fail to derive the original signal phase from Eq. (2.30). This problem is a key limitation of this method. Besides, if the target only performs tiny movements, i.e., the change of θ is small, then the change in phase difference may not show a clear movement pattern. What's more, this method requires the wireless system to be equipped with at least two receiving antennas, which sacrifices the limited number of antennas on existing wireless devices. Hence, it is more desirable to remove the phase noises of each individually received signal and obtain the original signal phase.

Therefore, we introduce several representative approaches to directly remove the phase noises. First, we show how to estimate the CFO by using the physical-layer signal. We set a calibration phase when the sensing environment is static without the target. Then, we acquire the IQ values of the down-converted received signal. If the CFO does not exist between the signal transmitter and receiver, i.e., $\Delta f = 0$, the received signal phase is relatively constant. If $\|\Delta f\| > 0$, the signal phase will change with a certain frequency, which is exactly Δf. Thus, by calculating the frequency of the signal phase of the received signal at the calibration phase, we can estimate Δf. Finally, we can perform a second-round down-conversion based on Δf and apply a low-pass filter to remove the $2\Delta f$ frequency component. The signal phase of the calibrated signal is the original phase without CFO.

Second, we show how to estimate and remove the CFO, SFO, and PDD if the wireless system adopts the OFDM [24]. Let us re-organize Eq. (2.29) as follows:

$$\varphi_m = \varphi_r + 2\pi(\sigma + \Delta f) \cdot f + \psi + Z. \tag{2.31}$$

As there are multiple subcarriers in OFDM, we can treat Eq. (2.32) as a function of the signal frequency f and represent it as

$$\varphi_m(f) = \varphi_r + af + b, \tag{2.32}$$

where $a = 2\pi(\sigma + \Delta f)$ and $b = \psi + Z$. Then, we can obtain a and b, which are the linear transformation coefficients, by solving the following optimization problem:

$$arg \min_{a,b} \sum_{i=1}^{I} (\varphi_m(f_i) - af_i - b)^2. \tag{2.33}$$

Theoretically, for an evenly distributed f in OFDM, a and b can be estimated as follows:

$$a = \frac{\varphi_m(f_I) - \varphi_m(f_1)}{f_I - f_1},$$

$$b = \frac{1}{I}\sum_{i=1}^{I}\varphi_m(f_i).$$

Consequently, we can remove a and b out of the signal phase.

The above method entails a critical assumption in which all sources of phase noises follow a linear relationship across different frequencies. However, the IQ imbalance will lead to non-linear phase errors [20]. The effect of IQ imbalance on the signal phase can be formulated as

$$\varphi_m(f_i) = atan(\epsilon_{i,a} \cdot \frac{sin(2\pi \Delta f + \epsilon_{i,\varphi})}{cos(2\pi \Delta f)}) - 2\pi f_i \sigma, \tag{2.34}$$

where $\epsilon_{i,a}$ and $\epsilon_{i,\varphi}$ refer to the gain and phase errors for the frequency f_i owing to the IQ imbalance, respectively. σ is the timing offset caused by PDD and SFO. Fortunately, the phase errors incurred by the IQ imbalance are relatively stable. As such, we can collect the signal in the calibration phase to estimate $\epsilon_{i,a}$ and $\epsilon_{i,\varphi}$ based on Eq. (2.34). With a sufficient number of signal points, least-square regression analysis can be employed to estimate $\epsilon_{i,a}$ and $\epsilon_{i,\varphi}$, which are used to remove the non-linear phase error.

Finally, let us discuss the frequency response of wireless devices. The performance of many wireless systems is related to their devices' frequency responses, such as, the transmitting and receiving antenna, speaker, and microphone. For RF systems, the antenna is designed to have a high and flat frequency response of a certain frequency or a specified frequency band. For example, the frequency response of NFC antennas is configured to reach the highest for their resonant frequency at 13.56 MHz. The ideal frequency response should be flat for wideband RF systems. The same is true for acoustic devices. The high-end speaker and microphone usually have a stable and flat frequency response across the whole frequency band covering 20 Hz to 20 KHz. As shown in Fig. 2.17a, the ideal frequency response should be flat. However, given the manufacturing imperfections

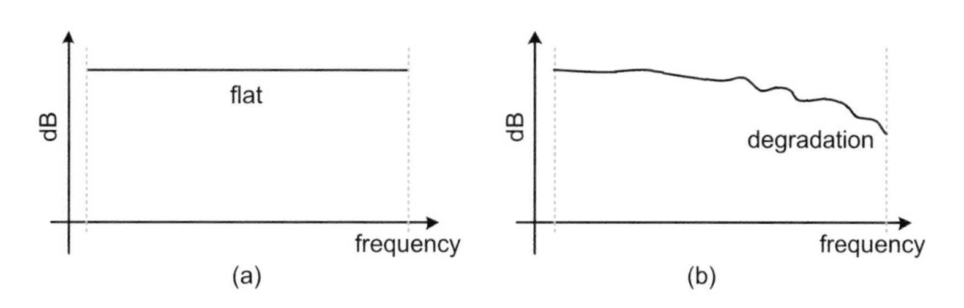

Fig. 2.17 Frequency response of wideband wireless device: (**a**) ideal flat frequency response, (**b**) actual frequency response with degradation

of the device hardware, the actual frequency response is degraded, as depicted in Fig. 2.17b. A degraded frequency response can adversely affect the signal amplitude and phase across the frequency band. Furthermore, different wireless devices suffer from various degradations of the frequency response, which means that the frequency response will change from one device to another. This phenomenon can lead to unstable sensing performance due to the device heterogeneity.

To calibrate the frequency response, we need to measure the frequency response in advance and then compensate for its effect. Many delicate devices can be used to measure the frequency response of antennas or acoustic devices. They usually require a special environment, such as the anechoic room and chamber room, to avoid the multipath effect for precise frequency response measurement [25]. However, these specialized devices are expensive, and it is cumbersome to set up a special environment.

Many software programs have been developed to ease the requirements for measuring the frequency response. For example, ARTA, which is a loudspeaker measurement program, was designed for audio measurements [26]. It was developed by Ivo Mateljan of the Electroacoustic Laboratory in Split, Croatia. The advantage of ARTA is that it does not require a special room. Moreover, the measurement is independent of the room response. After acquiring the device frequency response, we can design a corresponding filter whose frequency response is the reciprocal of the device's response [27]. We can utilize this filter to calibrate the non-flat frequency response.

2.3.2 Signal Processing to Release Signal Constraints

Apart from signal noises caused by imperfect wireless devices, the constrained resources of wireless devices also limit the sensing performance. One of the most important resources is the number of transmitters and receivers.[7] However, most existing wireless devices have a limited number of transmitters and receivers. Similarly, the intrinsic properties of certain wireless signals, including the signal carrier frequency, power, and bandwidth, also affect the sensing capability. This section will discuss the impacts of limited number of transmitters and receivers, signal carrier frequency, signal power and its attenuation, and signal bandwidth on the sensing performance and capability.

[7] For RF systems, the number of transmitting and receiving antennas is a key consideration. Existing wireless communication systems are equipped with multiple Tx and Rx antennas to improve the throughput.

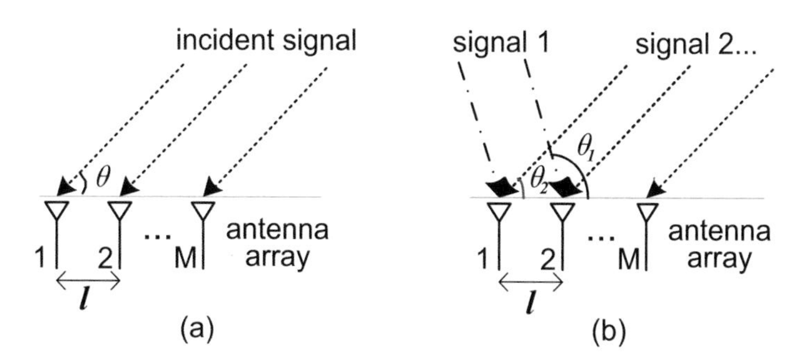

Fig. 2.18 AOA estimation with antenna array: (**a**) single incident signal, (**b**) multiple incident signals

2.3.2.1 Limited Number of Transmitters and Receivers

The measurement of the signal AOA, which is a key indicator for spatial sensing, requires multiple receiving antennas. In AOA estimation, the resolution highly relies on the number of antennas. Let us briefly introduce the AOA estimation using the antenna array, as shown in Fig. 2.18a. The AOA θ is given by $\theta = cos^{-1}(\frac{\lambda \Delta\varphi}{2\pi l})$, where $\Delta\varphi$ is the phase difference between two adjacent antennas.[8] While, the accuracy of θ is related to its resolution. Suppose there are two incident signals whose AOAs are θ and $\theta + \Delta\theta$, respectively. Then, the phase difference of the first incident signal is $\Delta\varphi_1 = \frac{2\pi}{\lambda} \cdot l \cdot cos(\theta)$; similarly, for the second incident signal is $\Delta\varphi_2 = \frac{2\pi}{\lambda} \cdot l \cdot cos(\theta + \Delta\theta)$. When we subtract $\Delta\varphi_2$ from $\Delta\varphi_1$, we obtain $\Delta\omega$ as follows:

$$\Delta\omega = \Delta\varphi_2 - \Delta\varphi_1 = \frac{2\pi l}{\lambda} \cdot [cos(\theta + \Delta\theta) - cos(\theta)]. \tag{2.35}$$

As $cos(\theta)$ is the derivative of $cos(\theta)$, $cos(\theta + \Delta\theta) - cos(\theta)$ can be approximated as $cos(\theta) \cdot \Delta\theta$. Then, $\Delta\omega$ becomes

$$\Delta\omega = \frac{2\pi l}{\lambda} \cdot cos(\theta) \cdot \Delta\theta. \tag{2.36}$$

To calculate the AOA, we can perform FFT on the signal phase among multiple antennas. The FFT is the M-point FFT, where M is the number of antennas. A peak will be observed in the FFT result for the AOA angle. Suppose that the abovementioned two incident signals will have distinct peaks in the FFT result, which means their peaks are more than $2\pi/M$ away from each other, i.e., $\Delta\omega > 2\pi/M$. By taking Eq. (2.36) into this inequality, we have:

[8] We will introduce details for AOA estimation in Sect. 4.2.1.2.

$$\frac{2\pi l}{\lambda} \cdot cos(\theta) \cdot \Delta\theta > \frac{2\pi}{M}, \ \Delta\theta > \frac{\lambda}{M \cdot l \cdot cos(\theta)} \tag{2.37}$$

As such, the AOA resolution $\Delta\theta$ is jointly affected by the number of antennas M and the AOA θ itself. When $\theta = 0$ and $l = \lambda/2$, we can yield the best AOA resolution, which is $2/M$. Even though there are advanced algorithms to upgrade the AOA resolution, e.g., MUltiple SIgnal Classification (MUSIC) [28, 29], we can always obtain a higher AOA resolution with more receivers/transmitters. We simulate the MUSIC algorithm, which calculates a spectrum of different AOAs, by using different numbers of antennas. The corresponding peak in the spectrum indicates the estimated AOA. We create two signals whose AOAs are 19° and 22°, respectively, and combine them. Then, we choose four and eight antennas to execute the MUSIC algorithm. As shown in Fig. 2.19a, when four antennas are used, the peak at 19° is degraded. By contrast, when the number of antennas is eight, there are two clear peaks at both 19° and 22°.

In practice, the number of incident signals can easily exceed two because there are many objects and reflectors in real environments. In a typical indoor environment, the number of effective signal paths is usually six to eight. However, most of the existing commodity wireless devices, e.g., WiFi router, network interface card, smart microphone, and radar antenna, are equipped with less than four antennas/receivers. The limited number of receivers can lead to an inaccurate estimation of the AOA in practical environments and degrade the performance of many applications requiring the spatial information, including indoor localization and gesture tracking.

A virtual antenna array is created to tackle the challenge of a limited number of physical antennas either by (1) moving the antenna or (2) employing the target's movement. The first method is to move the antenna to emulate an antenna array in the air [30, 31]. As shown in Fig. 2.20, when moving/rotating the antennas, a linear/circular antenna array is formed. The virtual antenna array not only expands the number of antennas but also provides an opportunity for removing phase errors. Suppose the phases of the signal from the antenna at two locations are φ_1 and φ_2, and the corresponding AOAs are θ_1 and θ_2, respectively. On the basis of Eq. (2.29),

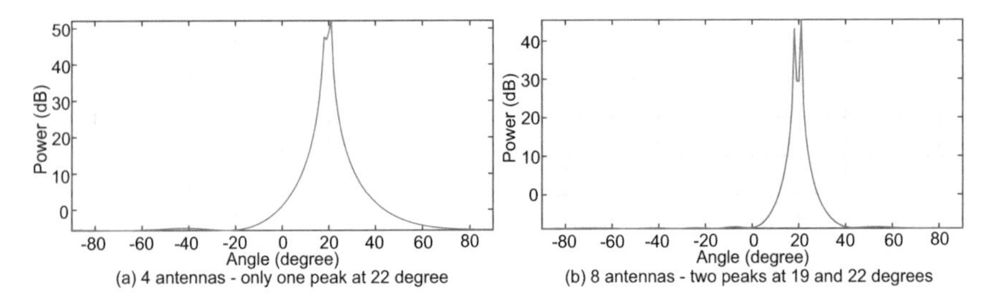

Fig. 2.19 Estimation of AOAs of two incident signals (19° and 22°) using different number of antennas by the MUSIC algorithm: (**a**) four antennas with only one peak at 22°, (**b**) eight antennas with two peaks at 19° and 22°, respectively

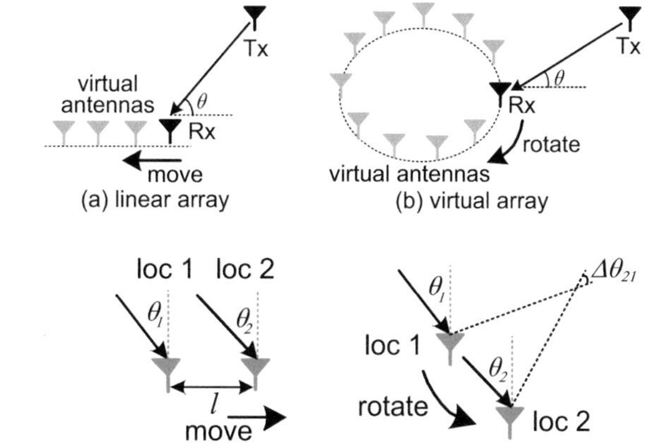

Fig. 2.20 Creation of virtual antennas: (**a**) linear antenna array, (**b**) circular antenna array

Fig. 2.21 AOA of signal when the antenna moves or rotates from location 1 (loc 1) to location 2 (loc 2)

the phase difference between locations 1 and 2, as depicted in Fig. 2.21, can be expressed as follows:

$$\varphi_{21} = \varphi_2 - \varphi_1 = 2\pi \cdot \frac{l}{\lambda} \cdot (sin\theta_2 - sin\theta_1). \tag{2.38}$$

As the phase errors are fixed for a single antenna, the phase errors are eliminated in the phase difference of two antenna locations. Then, we define an angle $\theta_{21} = arcsin(sin\theta_2 - sin\theta_1)$. We can also apply the classical AOA estimation, e.g., MUltiple SIgnal Classification (MUSIC) algorithm, to obtain θ_{21}. In the meantime, we can employ inertial sensors, i.e., gyroscope, to measure how the antenna is moved or rotated and calculate $\Delta\theta = \theta_2 - \theta_1$. In summary, we have the following two variables θ_2 and θ_1 and two equations:

$$\begin{cases} \theta_{21} = arcsin(sin\theta_2 - sin\theta_1) \\ \Delta\theta = \theta_2 - \theta_1 \end{cases} \tag{2.39}$$

After measuring θ_{21} and $\Delta\theta$, we can derive the AOA using the virtual antenna array. The accuracy of AOA can be further enhanced by employing multiple measurements from different antenna locations and aggregating multiple estimated AOAs.

In addition to moving the antenna, the movement of the target can also enable the formulation of an antenna array, which is called inverse synthetic aperture radar (ISAR). As illustrated in Fig. 2.22, when the target moves to the right with multiple signal measurements obtained from consecutive timestamps, it can be inversely considered as the target being static, but the antenna is moving. Suppose we have k consecutive measurements of the signal channel response, i.e., $h[n], h[n+1], \ldots, h[n+k]$, we can traverse all candidate AOAs and compute the spatial spectrum $A[\theta, n]$ as follows [32]:

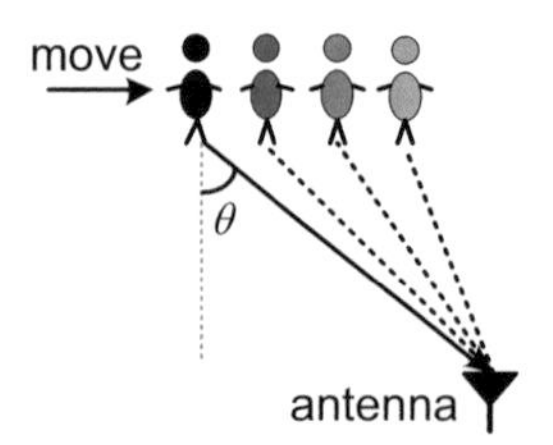

Fig. 2.22 Virtual antenna array emulated by the moving target

$$A[\theta, n] = \sum_{i=0}^{k} h[n+i]e^{j\frac{2\pi}{\lambda}\cdot i\cdot \Delta l\cdot sin\theta}, \tag{2.40}$$

where Δl is the distance of the target moving between two adjacent measurements. If we know the moving speed v_t of the target and $\Delta l = v_t \cdot \Delta t$, where Δt is the sampling interval, then we can obtain $A[\theta, n]$ for each timestamp. Moreover, $A[\theta, n]$ will reach the highest value when θ is traversed to the corresponding AOA of the target. The abovementioned methods emulate the virtual antenna array via the moving antenna/target. However, in some cases, it is cumbersome to make the antenna move/rotate, and the target is stationary. In these circumstances, these methods may not be effective.

There is a method that can create a virtual antenna array without moving the antenna/target. The key idea is to employ multiple subcarriers with different signal frequencies as virtual antennas. In specific, suppose the signal is transmitted on multiple subcarriers (the number of subcarriers is denoted as m), and the signal frequencies of each subcarrier are usually evenly distributed, i.e., $f_1, f_2, \ldots, f_m$ (the interval between adjacent frequencies is denoted as Δf), the corresponding signal phase for the signal path with a time delay of τ for those subcarriers can be represented as:

$$\varphi_i = \varphi_1 + 2\pi \cdot (i-1) \cdot \Delta f \cdot \tau. \tag{2.41}$$

Recall that for the physical linear antenna array with l distance away from each other, the signal phase of antenna j is expressed as follows:

$$\varphi_j = \varphi_1 + 2\pi \cdot (j-1) \cdot l \cdot sin\theta. \tag{2.42}$$

A comparison between Eqs. (2.41) and (2.42) indicates their similarity in which Δf plays the same role of l. Based on this insight, the subcarriers can be employed as additional antennas [33]. Take the WiFi network interface card with three antennas and 30 subcarriers as an example. The number of "antennas" after expanding can reach as high as 90, which can significantly promote the AOA resolution.

2.3.2.2 Effects of Signal Frequency and Power

The signal wavelength λ limits the resolution of measuring the change of signal propagation distance Δd from the signal phase φ, i.e.,

$$\varphi = \{\frac{d_0 + \Delta d}{\lambda}\} \mod 2\pi. \tag{2.43}$$

For a fixed Δd, the smaller is the signal wavelength, the more change that φ experiences. The signal wavelength is jointly decided by the signal frequency and propagation speed in the medium. Since the signal propagation speed is fixed in a given medium, the sensitivity of the wireless signal to the target's movement is mainly determined by the signal frequency. In other words, a high signal frequency is more capable of measuring the tiny propagation path change caused by the target. Thus, signals with smaller wavelengths (e.g., 60 GHz WiFi and mmWave radar signals) are usually leveraged to measure tiny movements, such as machine vibrations and human vital signs. For signals with large wavelengths (e.g., Lora, RFID, and 2.4 GHz WiFi signals), they are less sensitive to the tiny target movements.

In real sensing environments, the displacement resolution is not only limited by the signal wavelength because the received signal is a superimposition of multiple signal paths, including static signals reflected by surrounding stationary objects (H_s) and the dynamic signal reflected by the moving target (H_d), as shown in Fig. 2.23. The real signal phase change is φ_r. However, in the presence of H_s, the measured signal phase becomes φ_m, which is smaller than φ_r. This condition means that the sensitivity of displacement measurement is decreased. Thus, the effect of H_s should be removed properly.

One way to eliminate H_s is to move the original point to the start of H_d, i.e., point A, by subtracting the average of the complex signal from the original signal. For some tiny movements, we can further enlarge the phase change to be larger than the real one. To achieve this goal, researchers have proposed to add a complementary signal vector based on two key observations [15]: (1) reducing the amplitude of

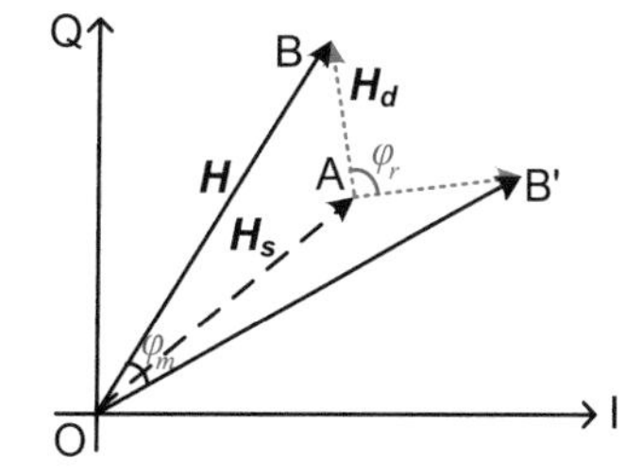

Fig. 2.23 Real and measured signal phase with the effect of static signal component, which the measured signal phase smaller than the real signal phase

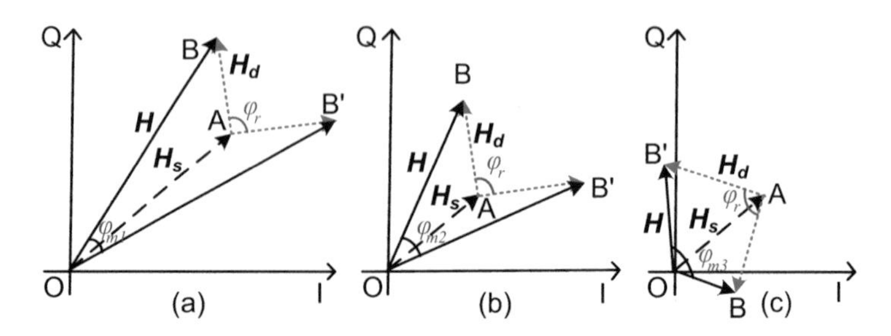

Fig. 2.24 Signal phase changes with different amplitudes of H_s and starting positions of H_d

static vector H_s can enlarge the signal phase change φ_m.[9] As shown in Fig. 2.24a and b, when the length H_s decreases, φ_m increases accordingly, i.e., $\varphi_{m2} > \varphi_{m1}$. (2) the starting position of H_d affects the size of φ_m. As shown in Fig. 2.24b and c, H_d starts from two different positions and incurs different phase changes, i.e., $\varphi_{m3} > \varphi_{m2}$. Therefore, we can create a complementary signal vector H_c and add it to the static vector H_s to produce a new static vector H_{sn}. The amplitude of H_{sn} should be smaller than that of H_d, and the largest φ_m can be acquired. In this way, we can exaggerate the displacement resolution.

A high signal frequency is helpful for sensing tiny movements, however, it also poses a problem. The high-frequency signal suffers from more power attenuation because it can be easily blocked by barriers given it small wavelength. When the signal's wavelength is less than the size of surrounding objects, the signal will be reflected. The diffracted effect of the signal can also decrease. Take the case when you connect to the WiFi at the 5 GHz frequency band. The signal strength quickly drops when you move far away. However, the signal strength of the 2.4 GHz frequency band is much higher than that of the 5 GHz frequency band. This is mainly because the signal wavelength of 5 GHz is around two times smaller than that of 2.4 GHz. Simply increasing the signal power or the antenna gain may not properly solve this issue because the Federal Communications Commission (FCC) regulates the RF energy.

Wireless signals also manifest different features across a variety of channels. Acoustic signals attenuate less in liquids compared with RF signals. Light signals cannot go through walls. RF signals can "go through" walls owing to the signal diffraction. In general, most existing wireless sensing systems can work properly only within 10 m. Researchers are currently investigating solutions to extend the sensing range. To expand the sensing range, recently, many researchers employ LoRa to achieve long-range sensing. LoRa adopts the chirp spread spectrum (CSS) to encode the data. The reason why LoRa can work in the long range is because it introduces the spreading factor. The spreading factor reflects how many chirps are sent per second. A higher spreading factor means a sharper chirp in the same

[9] We use the IQ plane to model the effect of static and dynamic signals. Please refer to Sect. 4.2 for more details about signal modeling in the IQ plane.

frequency band. As the number of chirps decides the number of bits of data, a higher spreading factor leads to a higher data rate and vice versa. The most interesting part of LoRa is that it has different spreading factors to select.[10] For communication in the short range, high spreading factors are used with a high data rate. Meanwhile, in the long range, as a longer chirp can help to decode the data bit with higher accuracy, short spreading factors are employed, but with a low data rate. Therefore, there is a trade-off between communication range and data rate. Existing studies have shown that using LoRa can achieve over 50 m sensing range [34, 35].

2.3.2.3 Limited Bandwidth

In Sect. 2.3.2.2, we introduce that the signal phase can be used to measure the change of signal propagation distance. However, purely using the signal phase cannot report the absolute signal propagation distance between the sensing target and wireless transceivers. To derive the absolute signal propagation distance d, one solution is to obtain the signal propagation time τ of the signal, then $d = v \cdot \tau$, where v is the propagation speed of the signal. As v is a constant, the resolution of τ determines how precise d can be estimated. A more fine-grained τ can derive a more accurate estimation of d. In many wireless systems, the high resolution of time τ is achieved by expanding the bandwidth B of the signal because the temporal resolution depends on the width of the frequency band.[11] Specifically, the resolution of signal propagation time is given by $\tau_{res} = \frac{1}{B}$.

Narrowband signals, such as WiFi and LoRa, suffer from low accuracy during distance measurement. The bandwidth of the current 802.11 WiFi frequency band is 20/40 MHz. For the bandwidth of 40 MHz, $\Delta\tau$ is 25 ns. This scenario means that the resolution of the path length for the two signal paths is $\Delta\tau \times c = 7.5$ m. The newly developed WiFi 6 standard expands the bandwidth to 160 MHz. However, the distance resolution is still higher than 1.5 m. Such a resolution can lead to uncertainties in many sensing applications. For example, the error in estimating a person's location or gesture trajectory via TOF is at least 7.5 m, which is not acceptable for real use. For acoustic signals in the frequency band of 18–22 KHz,[12] $\Delta\tau$ is around 2.5×10^{-4} s. The corresponding distance resolution is around 8.6 cm, which is much smaller than that of the WiFi signal. This situation can be contributed to sound speed, which is much slower than electromagnetic signals. However, the highest frequency of most commodity speakers and microphones is under 24 KHz, which limits the distance resolution and causes it to stay at the centimeter level.

---

[10] The spreading factors are SF7, SF8, SF9, SF10, SF11, and SF12.

[11] When performing FFT on a signal, the frequency resolution in the FFT result is decided by the time span of the signal. Similarly, for IFFT, the temporal resolution is determined by the bandwidth.

[12] The 18–22 KHz frequency range is the commonly used frequency band in acoustic sensing using commodity speakers and microphones because the sound in this range is inaudible to people and cause less disturbance.

Wideband signals, e.g., ultrasound, UWB radar, and FMCW radar, can achieve higher resolutions when measuring the signal propagation delay. For the 4 GHz bandwidth radar, d_{res} is 3.75 cm, and the corresponding resolution for measuring the distance between the target and radar is only 1.875 cm. However, wideband wireless devices are more expensive and unavailable to the public, which hinders the wide usage for sensing. Therefore, it is a challenging task to achieve high-resolution sensing on narrowband wireless devices.

The main approach for broadening the bandwidth is to fully utilize the frequency band available on wireless devices. For WiFi devices, even though the bandwidth of each individual band is narrow (20/40 MHz), the whole bandwidth is actually more than tens of MHz, e.g., the total bandwidth allocated for 5 GHz WiFi is over 200 MHz. In addition, RFID also has several frequency bands. As such, we can perform frequency hopping across multiple bands to expand the bandwidth [36, 37]. The hopping delay among different frequency bands should be as small as possible to guarantee that the hopping can be finished within the coherence time.[13] For acoustic signals, the frequency band of 18–22 KHz is mainly utilized for acoustic sensing on commodity speakers and microphones because the sound in this band is mostly inaudible to people so that the acoustic sensing system will not cause disturbance to users. However, the limitation is that the bandwidth is only 4 KHz. To make use of the whole frequency band from 100 Hz to 22 KHz and reduce the sound disturbance, researchers have designed white noises that cover the band between 100 Hz and 22 KHz as the transmitted acoustic signal [38].

2.3.2.4 Effect of Frequency Selective Fading on Sensing

Frequency selective fading is a phenomenon of signal fading at certain frequencies due to the presence of multiple signal paths in wireless communication systems. This phenomenon is also harmful in wireless sensing. As shown in Fig. 2.25, there are two signal paths between the Tx and Rx. Signal path 1 is the one reflected by the target, while signal path 2 is an irrelevant path caused by a surrounding object. Then, the difference of the propagation time between path 1 and path 2 is given by $\Delta\tau$. If the signal frequency equals to $\frac{1}{2 \cdot \Delta\tau}$, then the received signal, which is the superposition of signal path 1 and path 2, would become zero. Because two signals

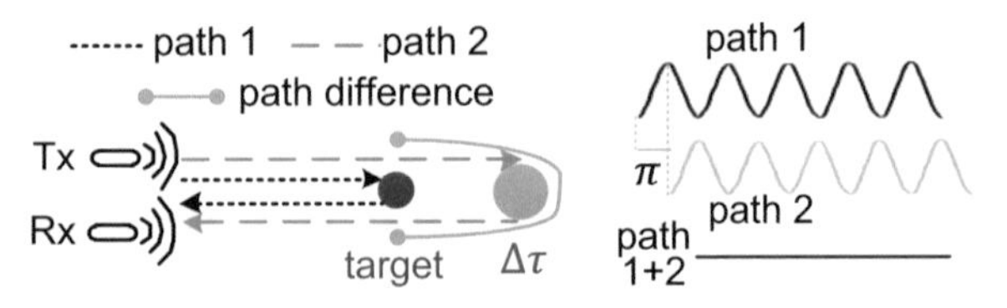

Fig. 2.25 Illustration of frequency selective fading phenomenon

[13] Coherence time is the time duration during which the channel impulse response does not change.

have opposite phases and destructively cancel each other. For other frequencies, the fading phenomenon will not appear. Thereby, it is called frequency selective fading.

For the purpose of sensing under a given signal frequency, if the target is unexpectedly located at some places where signal fading happens, we are unable to detect the target. The solution to this problem is to adopt the frequency hopping to transmit the signal in multiple frequencies. In this way, we can ignore the fading effect from the incurring frequency and maintain the signal pattern of the target through "safe" frequencies [3].

References

1. Molisch AF (2012) Wireless communications, vol 34. John Wiley & Sons
2. Wang W et al (2016) Device-free gesture tracking using acoustic signals. In: Proceedings of the 22nd annual international conference on mobile computing and networking, pp 82–94
3. Wang Y et al (2020) Push the limit of acoustic gesture recognition. IEEE Trans Mob Comput 21(5):1798–1811
4. Zheng T et al (2020) V2iFi: in-vehicle vital sign monitoring via compact rf sensing. Proc ACM Interactive Mob Wearable Ubiquitous Technol 4(2):1–27
5. Cheng H, Lou W (2021) Push the limit of device-free acoustic sensing on commercial mobile devices. In: Proceedings of IEEE conference on computer communications, pp 1–10
6. Zepernick HJ, Adolf F (2013) Pseudo random signal processing: theory and application. John Wiley & Sons
7. Haim A (2010) Basics of biomedical ultrasound for engineers. John Wiley & Sons
8. Cai C et al (2020) AcuTe: acoustic thermometer empowered by a single smartphone. In: Proceedings of the 18th conference on embedded networked sensor systems, pp 28–41
9. Finkenzeller K (2010) RFID handbook: fundamentals and applications in contactless smart cards, radio frequency identification and near-field communication. John Wiley & Sons
10. Wang H et al (2016) Human respiration detection with commodity wifi devices: do user location and body orientation matter? In: Proceedings of the ACM international joint conference on pervasive and ubiquitous computing, pp 25–36
11. Zhang F et al (2018) From fresnel diffraction model to fine-grained human respiration sensing with commodity wi-fi devices. Proc ACM Interactive Mob Wearable Ubiquitous Technol 2(1):1–23
12. Yang Z et al (2013) From RSSI to CSI: Indoor localization via channel response. ACM Comput Surv 46(2):1–32
13. Zhang F et al (2020) Exploring lora for long-range through-wall sensing. Proc ACM Interactive Mob Wearable Ubiquitous Technol 4(2):1–27
14. Zhang F et al (2021) Unlocking the beamforming potential of LoRa for long-range multi-target respiration sensing. Proc ACM Interactive Mob Wearable Ubiquitous Technol 5(2):1–25
15. Xie B et al (2021) Pushing the limits of long range wireless sensing with LoRa. Proc ACM Interactive Mob Wearable Ubiquitous Technol 5(3):1–21
16. Wang S et al (2020) Demystifying millimeter-wave V2X: Towards robust and efficient directional connectivity under high mobility. In: Proceedings of the 26th annual international conference on mobile computing and networking
17. Ha U et al (2020) Contactless seismocardiography via deep learning radars. In: Proceedings of the 26th annual international conference on mobile computing and networking, pp 1–14
18. Li T et al (2015) Human sensing using visible light communication. Proceedings of the 21st annual international conference on mobile computing and networking, pp 331–344

19. Ma D et al (2019) Solargest: Ubiquitous and battery-free gesture recognition using solar cells. In: The 25th annual international conference on mobile computing and networking, pp 1–15
20. Zhou Y et al (2017) Perceiving accurate CSI phases with commodity WiFi devices. In: IEEE conference on computer communications, pp 1–9
21. Zhuo Y et al (2017) Perceiving accurate CSI phases with commodity WiFi devices. In: IEEE conference on computer communications, pp 1–9
22. Yang Y et al (2018) Wi-count: Passing people counting with COTS WiFi devices. In: 27th International conference on computer communication and networks, pp 1–9
23. Wang X et al (2017) PhaseBeat: Exploiting CSI phase data for vital sign monitoring with commodity WiFi devices. In: IEEE 37th international conference on distributed computing systems, pp 1230–1239
24. Wang G et al (2021) Dynamic phase calibration method for CSI-based indoor positioning. In: IEEE 11th annual computing and communication workshop and conference, pp 0108–0113
25. Foley JT et al (2017) Low-cost antenna positioning system designed with axiomatic design. In MATEC web of conferences, vol 127. EDP Sciences, p 01015
26. Ivo Mateljan. Making gated-impulse frequency measurements using ARTA. http://marjan.fesb.hr/~mateljan/
27. Mao W et al (2017) Indoor follow me drone. In: Proceedings of the 15th annual international conference on mobile systems, applications, and services, pp 345–358
28. Shahi SN et al (2008) High resolution DOA estimation in fully coherent environments. Prog Electromagn Res C 5:135–148
29. Li X et al (2016) Dynamic-music: accurate device-free indoor localization. In: Proceedings of the 2016 ACM international joint conference on pervasive and ubiquitous computing, pp 196–207
30. Qian K et al (2017) Enabling phased array signal processing for mobile WiFi devices. IEEE Trans Mob Comput 17(8):1820–1833
31. Kumar S et al (2014) Accurate indoor localization with zero start-up cost. In: Proceedings of the 20th annual international conference on mobile computing and networking, pp 483–494
32. Adib F, Katabi D (2013) See through walls with WiFi! In: Proceedings of the ACM SIGCOMM 2013 conference on SIGCOMM, pp 75–86
33. Kotaru M et al (2015) Spotfi: Decimeter level localization using wifi. In: Proceedings of the ACM conference on special interest group on data communication, pp 269–282
34. Zhang F et al (2020) Exploring LoRa for long-range through-wall sensing. Proc ACM Interactive Mob Wearable Ubiquitous Technol 4(2):1–27
35. Zhang F et al (2021) Unlocking the beamforming potential of LoRa for long-range multi-target respiration sensing. Proceedings of the ACM on Interactive, Mobile, Wearable and Ubiquitous Technologies 5(2):1–25
36. Tan S et al (2019) MultiTrack: Multi-user tracking and activity recognition using commodity WiFi. In: Proceedings of the 2019 CHI conference on human factors in computing systems, pp 1–12
37. Xie Y et al (2018) Precise power delay profiling with commodity Wi-Fi. IEEE Trans Mob Comput 18(6):1342–1355
38. Cai C et al (2021) Active acoustic sensing for hearing temperature under acoustic interference. IEEE Trans Mob Comput. Early Access

Chapter 3
Wireless Sensing System Configurations

3.1 Single-Transceiver vs. Multi-Transceiver Configurations

The single-transceiver setting denotes that the system only has one transmitter and one receiver. In the early stage of wireless sensing, the single-transceiver setting was mainly adopted because most devices were only equipped with a single transmitter or receiver [1, 2]. As shown in Fig. 3.1a, acoustic sensing systems use the smartphone with one pair of bottom microphone and speaker to send and receive the signal, respectively. In addition, in many RF-based systems, only one pair of transmitting and receiving antenna is employed, e.g., the single-antenna WiFi router and network interface card (NIC) shown in Fig. 3.1b and the single-transceiver UWB radar in Fig. 3.1c. Therefore, single-transceiver configurations are cheap and much easier to deploy without considering the time synchronization problem between multiple transmitters and receivers.

With the development of wireless devices, the multi-transceiver design has become a popular way for managing wireless sensing systems. Many wireless devices have since been equipped with multiple transmitters and receivers [3, 4]. As shown in Fig. 3.1d, four microphones are installed on the voice assistant device. The number of antennas on the WiFi router and NIC also increases from one to three or even eight, as depicted in Fig. 3.1e. The recently compact design of FMCW radar using mmWave also embed three transmitting antennas and four receiving antennas, as shown in Fig. 3.1f.

The original objective of the multi-transceiver design is to increase the communication throughput or reduce the effect of noises in the wireless channel. The surprise is that it also brings prominent benefits for sensing tasks. First, the usage of multiple receivers can help to eliminate noises in terms of both hardware imperfections and environment noises. Because these receivers are affected by the same source of noises. Therefore, the signal difference between two receivers can cancel out these noises. Second, multiple transceivers can improve spatial sensing capability. Similar

J. Cao, Y. Yang, *Wireless Sensing*, Wireless Networks,
https://doi.org/10.1007/978-3-031-08345-7_3

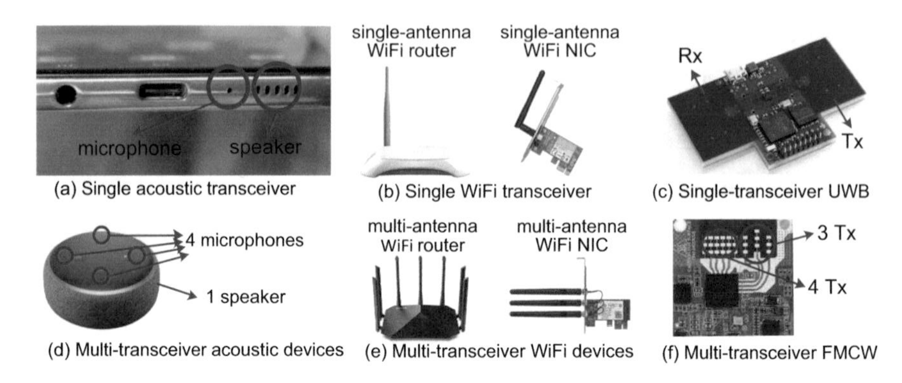

Fig. 3.1 Single-transceiver and multi-transceiver configurations and examples: (**a**) smartphone with a single bottom speaker and microphone, which act as the acoustic transmitter and receiver, respectively, (**b**) WiFi router and NIC with a single antenna, (**c**) UWB radar with one transmitter and receiver, (**d**) voice assistant with one speaker and four microphones, (**e**) WiFi router and NIC with multiple antennas, (**f**) FMCW radar with three transmitters and four receivers

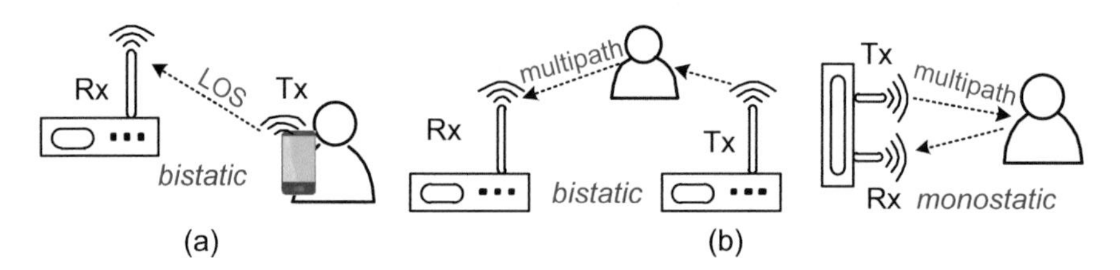

Fig. 3.2 (**a**) Device-based and (**b**) device-free system configurations

to human beings who have two ears to locate the sound source, the use of multiple transceivers enables the accurate localization of the target in space.

3.2 Device-Based vs. Device-Free Configurations

There are mainly two ways to deploy and place the wireless transmitter and receiver for sensing. The first way is to attach the wireless transmitter or receiver onto the sensing target, which is called device-based configuration. As shown in Fig. 3.2a, the target person carries a smartphone acting as the transmitter to send the signal to the remote receiver. The smartphone moves along with the target person. In this manner, the LOS signal between the smartphone and the receiver mainly involves the movement information. By deriving the pattern in the signal from the smartphone, we can infer the person's activities. The setup in Fig. 3.2a in which the transmitter and receiver are separately placed at a certain distance is also denoted as the bistatic setting.

In the second approach, the sensing target is freed from carrying any wireless device, which is called device-free configuration [5]. It releases the target from wearing the wireless transmitter or receiver, which provides a more convenient alternative for sensing. As shown in Fig. 3.2b, even though the target does not wear any device, it still can be sensed because the target's movements affect the multipath signal between the transmitter and receiver. The target, which acts as a reflector, affects signal propagation paths in various means when it moves. In the device-free configuration, the transmitter and receiver are either bistatic or monostatic. The monostatic setting refers to the collocated transmitter and receiver, as shown by the right panel in Fig. 3.2b. By analyzing the reflected multipath signal from the target, we can then obtain information about the target's movement.

The trend in wireless sensing currently is moving from the single-transceiver to multi-transceiver setting and from device-based to device-free configuration with higher sensing accuracy and better user experience. In addition, the monostatic setting is also attractive because it eases the setting of separate transceivers. For example, existing mmWave transceivers are mainly installed on a single small board, which is a portable setup.

References

1. Gong W, Liu J (2018) SiFi: Pushing the limit of time-based WiFi localization using a single commodity access point. Proc ACM Interactive Mob Wearable Ubiquitous Technol 2(1):1–21
2. Abdelnasser H et al (2015) Wigest: A ubiquitous wifi-based gesture recognition system. In: IEEE conference on computer communications, pp 1472–1480
3. Sun K et al (2018) Vskin: Sensing touch gestures on surfaces of mobile devices using acoustic signals. In Proceedings of the 24th annual international conference on mobile computing and networking, pp 591–605
4. Ren Y, Yang J (2021) 3D Human pose estimation for free-form activity using WiFi signals. Proc ACM Interactive Mob Wearable Ubiquitous Technol 5(4):1–29
5. Wu D et al (2017) Device-free WiFi human sensing: From pattern-based to model-based approaches. IEEE Commun Mag 55(10):91–97

Chapter 4
Wireless Sensing Methodologies

4.1 Sensed Information from Wireless Signals

In this section, we first present the underlying phenomenons behind the extraction of sensing target information from wireless signals.

We start from a simple bistatic and device-based system configuration in which only one transmitter and receiver are deployed in a bistatic way, and the receiver Rx is carried by the sensing target, as shown in Fig. 4.1a. If the target (i.e., Rx) is stationary, we can infer the static information of the target, e.g., distance, angle, and channel (i.e., the medium) between Tx and target from the line-of-sight (LOS) signal. If the target (i.e., Rx) is moving, for instance, moving from position (1) to position (2), as shown in Fig. 4.1a. The received LOS signal vector will change and rotate in the IQ plane accordingly, as depicted in Fig. 4.1b. The length of the signal vector is the signal amplitude, and the angle between the vector and the I-axis is the signal phase. Then, from the change of the signal vector, we can extract the moving information, e.g., moving speed, moving distance, and moving pattern (walking or run).

The above modeled phenomenon is ideal and impractical because the real sensing environment can be surrounded by many non-target objects. As shown in Fig. 4.2a, the surrounding objects can incur multipath signals. Then, another signal vector, which refers to the multipath signal, is added to the IQ plane in Fig. 4.2b. If the object is stationary, then the multipath signal vector is static. If the object also moves, then the multipath signal vector will also rotate together with the LOS signal. As such, the received signal is not only affected by the LOS signal but also the unwanted multipath signals. Therefore, we need to properly deal with the disturbances caused by unwanted multipath signals for accurate extraction of the sensing target's information.

Second, we model the signal propagation for the bistatic and device-free configuration. As shown in Fig. 4.3a, the received signal is a combination of the LOS signal between the Tx and Rx and the multipath signal reflected by the target.

J. Cao, Y. Yang, *Wireless Sensing*, Wireless Networks,
https://doi.org/10.1007/978-3-031-08345-7_4

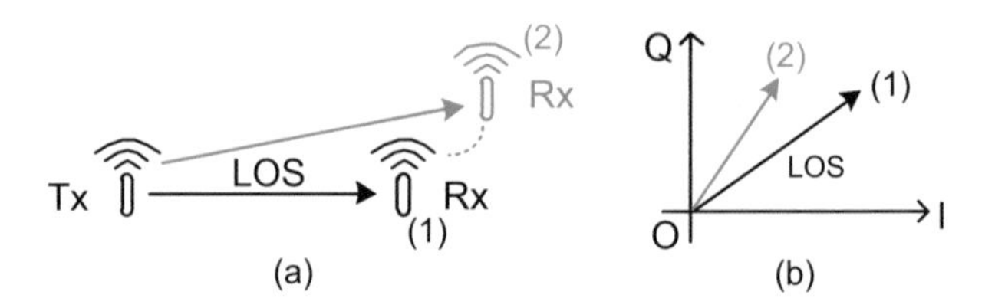

Fig. 4.1 Modeling the propagation of wireless signal when there is only one transmitter and receiver. (**a**) Signal propagation. (**b**) IQ plane

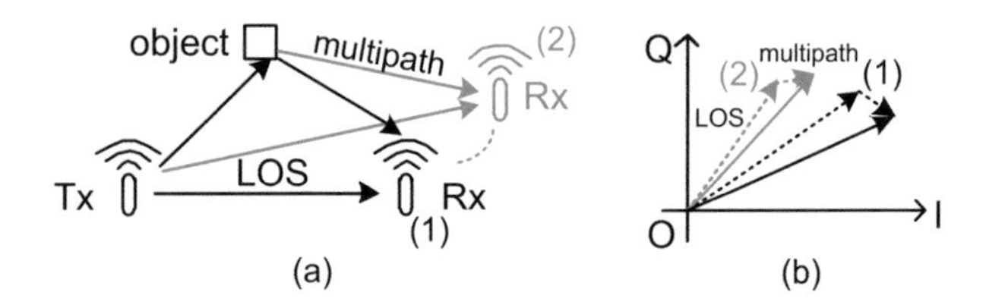

Fig. 4.2 Modeling the propagation of wireless signal when there are one transmitter, one receiver, and one object. (**a**) Signal propagation. (**b**) IQ plane

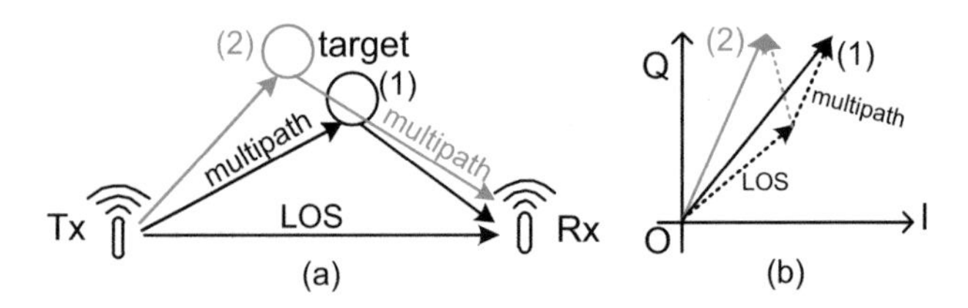

Fig. 4.3 Modeling the propagation of wireless signal for bistatic and device-free configuration: one pair of transceiver and one moving target. (**a**) Signal propagation. (**b**) IQ plane

If there is only one stationary sensing target in the environment, we can infer the target's information after removing the LOS signal. If the target moves to position (2), the LOS stay unchanged, but the multipath signal vector will rotate in the IQ plane and generate a new signal vector, as depicted in Fig. 4.3b. We can derive the target moving information and pattern from the change of multipath signal. For more practical scenarios, there can be other objects in the environment, which add other multipath signal vectors in the received signals. Therefore, we need to remove the effect from both the LOS signal and other non-target multipath signals for explicitly obtaining the target's information.

Third, we discuss the signal propagation for the monostatic and device-free configuration, as depicted in Fig. 4.4. The present target and non-target object both introduce multipath signals. For the stationary object in Fig. 4.4a with a static multipath signal, the signal vector caused by the object is fixed in Fig. 4.4b. Whereas the moving target results in a dynamic signal vector, which rotates from vector (1) to vector (2) in Fig. 4.4b. Accordingly, the overall received signal, which is the summation of static and dynamic multipath signals, rotates when the target moves.

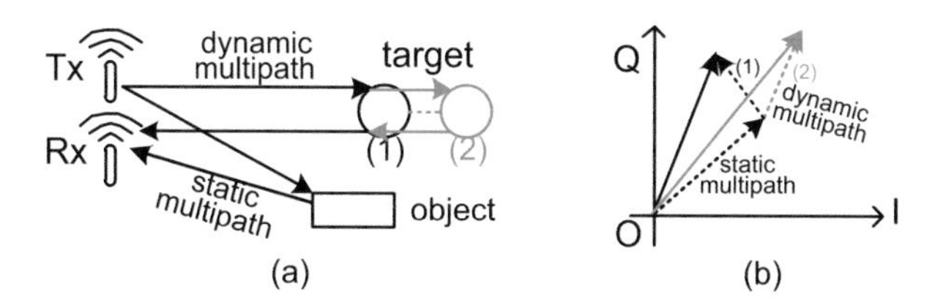

Fig. 4.4 Modeling the propagation of wireless signal for monostatic and device-free configuration: one pair of transceiver, one stationary object, and one moving target. (**a**) Signal propagation. (**b**) IQ plane

The target information is contained in the dynamic multipath signal. However, if both the target and object are moving, all the signal vectors will be dynamic. We need to separate the dynamic multipath signal reflected by the target from that reflected by the moving object for extracting the target's information.

Understanding each signal propagation path is an essential step for extracting the target's information. For the signal propagation path of the sensing target, it has the signal path length d, the signal angle of arrival (AOA), the signal amplitude a and the signal phase φ, from which we can derive the information for the sensing target.

There are mainly two categories of information that can be sensed from wireless signals, including physical information (e.g., target moving distance and speed) and semantic information (e.g., walking, sitting, and sleeping). For example, from the change of signal path length d when the target is moving, the physical information, such as, the moving speed v_m or frequency f_m, can be derived. These physical indicators are mainly obtained based on signal propagation laws and geometric relationships. On the basis of these physical indicators, we can also infer the semantic information, such as the type of activity the person is doing and the shape of the object. Apart from directly estimating the physical information from wireless signals, we can also extract features, which can uniquely represent the target's different states or movements, from wireless signals. We can handcraft and select some statistical features or adopt neural network models to automatically extract features. In the following content, we will introduce two main methodologies to obtain the target's information from wireless signals.

4.2 Model-Based Methodologies

Many wireless sensing applications are based on discovering the physical information from wireless signals. In this section, we introduce and discuss the model-based methodologies for obtain the target's physical information.

4.2.1 Wireless Signal Propagation Models for Sensing

There are many widely used physical indicators for sensing, and they can be derived from wireless signal propagation models that obey certain physical laws. We will show how to estimate physical indicators, including the signal propagation distance, the signal AOA between the sensing target and the wireless device, the speed, and the frequency of the moving target.

4.2.1.1 Signal Propagation Distance

The signal propagation distance d, which is a key indicator for sensing, can be estimated in different ways.

First, if we obtain the propagation time τ of the signal, then we can estimate the propagation distance via $d = \tau \times v$. UWB and FMCW radar are designed to derive τ between the radar antenna and the target. For the UWB radar, it repeatedly transmits a pulse signal $s(t)$, and $s(t - nT_s) = s(t)$, where n and T_s refer to the nth pulse and duration of the pulse. When the signal is reflected by the target, there is a time delay τ of the signal. τ can be derived in each received pulse signal by measuring how much time the echoed signal from the target is delayed from the pulse. The pulse is usually up-converted before sent out by the transmitter and down-converted after being received. The up-conversion and down-conversion can be specifically designed according to the application requirement [1].

For the FMCW radar, the frequency shift between the transmitted signal and received signal is employed to measure the signal propagation time τ. The FMCW radar transmits a chirp signal, whose frequency linearly increases over time, as shown in Fig. 4.5. Given the time delay τ between the transmitted chirp (Tx chirp) and received chirp (Rx chirp), the frequency of the signal also shifts by f_τ. During each chirp period (T_s), the Tx chirp and Rx chirp are combined by a "mixer" that produces an intermediate frequency (IF) signal. The frequency of the IF signal is denoted by f_τ. Therefore, by performing fast Fourier transformation (FFT) on the IF signal, we can estimate f_τ and calculate the propagation distance.

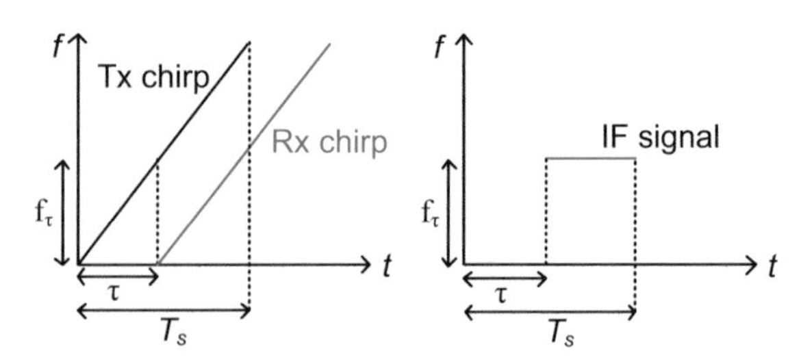

Fig. 4.5 Sweeping frequency of the FMCW radar signal and the IF signal after down-conversion

For UWB and FMCW radars, the resolution d_{res} for measuring the propagation distance depends on the bandwidth B, which is $d_{res} = v \cdot \frac{1}{\tau_{res}} = \frac{v}{B}$.[1] As B is usually in the gigahertz level, d_{res} is relatively small. For the narrowband signals, e.g., WiFi and RFID, this approach leads to a larger resolution.

If the transmitter and receiver are synchronized, e.g., pertaining to the speaker and microphone in the smartphone, then τ can also be estimated by calculating the correlation between the transmitted signal and received signal. For example, we can select a special code, e.g., the Zadoff-chu sequence with a high auto-correlation performance, as the transmitted signal. After calculating the cross-correlation between the transmitted $s(t)$ and received signal $r(t)$, i.e., $h(t) = r^*(t) \times s(t)$, we obtain a high peak at $t = \tau$. Then, τ can be obtained by extracting the location of the peak [2].

The second method is to estimate d from the received signal power P_r. For RF signals, P_r can be expressed based on the Friis equation as follows:

$$P_r = \frac{P_t G_t G_r \lambda}{(4\pi d)^2}, \tag{4.1}$$

where P_t is the power of the transmitter signal. G_t and G_r are the gains of the transmitter and receiver, respectively. However, the received signal is the superposition of multiple signals from different paths. P_r not only involves the signal from the target but also other surrounding objects. Hence, we may obtain an inaccurate d by using Eq. (4.1).

The third method is to leverage the signal phase φ to estimate the change of d, i.e., Δd. The signal phase φ can reflect the distance as follows:

$$\varphi = \frac{d_0 + \Delta d}{\lambda} \mod 2\pi, \tag{4.2}$$

where d_0 is the initial distance between the object and Tx/Rx. As φ is within $[0, 2\pi]$, the modulo operation is added. The phase needs to be unwrapped to show the total distance change. Then, φ will change based on Δd. Thus, Eq. (4.2) is mainly used for the moving target. In addition, the change of φ can provide a more fine-grained estimation of Δd. As explained previously for UWB and FMCW radars, the resolution of d_{res} relies on the bandwidth. Even for a wide bandwidth as 4 GHz and signal speed of 3×10^8 m/s, the d_{res} is around 3.75 cm. For smaller movements like respiration and heartbeat, the displacement is only at the millimeter level. Hence, d_{res} is not enough for observing the respiration activity. By contrast, φ can characterize Δd in a more sensitive way. For the signal with a frequency of 24 GHz whose wavelength is 12.5 mm, the change of φ can capture the millimeter-level change of Δd.

[1] The resolution of time τ_{res} is the inverse of bandwidth B.

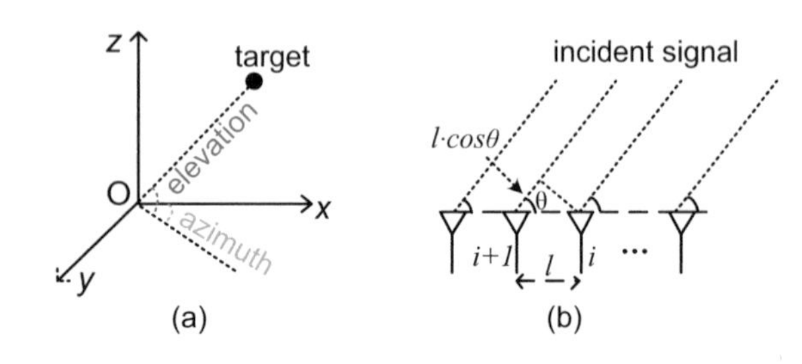

Fig. 4.6 AOA: (**a**) azimuth and elevation, (**b**) the antenna array to estimate AOA

4.2.1.2 Angle of Arrival

The signal AOA includes the azimuth and elevation angles, as illustrated in Fig. 4.6a. Taking the azimuth as an example, the AOA is mainly estimated by employing an array of antennas, as depicted in Fig. 4.6b. Suppose n antennas are placed with an equal interval l between each them of them.[2] Then, the signal arriving at the $(i+1)$th antenna, as opposed to the ith antenna, is a bit delayed. The delayed part can be described by the signal phase difference $\Delta\varphi$, which is expressed as:

$$\Delta\varphi = \frac{2\pi}{\lambda} \cdot l \cdot cos\theta. \tag{4.3}$$

By measuring the $\Delta\varphi$, we can infer the AOA as $\theta = cos^{-1}(\frac{\lambda\Delta\varphi}{2\pi l})$. Notably, $\Delta\varphi$ relies on $cos\theta$, which is called nonlinear dependency. $cos\theta$ is approximated with a linear function only when θ has a small value, i.e., $cos\theta \sim \theta$.

In practice, estimating AOA is challenging in real environments with multiple signal paths. Under this circumstance, we need to simultaneously estimate AOAs for all paths.

4.2.1.3 Moving Frequency and Speed

If the target moves repetitively with a certain frequency, we can estimate the moving frequency from both signal amplitude and phase. As shown in Fig. 4.7, the target's repetitive movement causes the dynamic vector H_d to shift back and forth from B and B'. H_s is the static vector. Then, the signal amplitude $\|H\|$ and the phase (the angle between OB/OB' and I-axis) will increase and decrease in a periodic manner. Therefore, the target's movement frequency can be estimated via FFT.

As illustrated in Eq. (4.2), the signal phase change can reflect the moving distance of the target. Then, by taking the derivative of phase $\frac{\Delta\varphi}{\Delta t} \sim \frac{\Delta d}{\Delta t} \sim v_m$, we may obtain the moving speed v_m. Moreover, v_m can be estimated using the channel response, e.g., the channel state information [3]. When there are many moving

[2] l is usually set to the half of wavelength to avoid phase ambiguity.

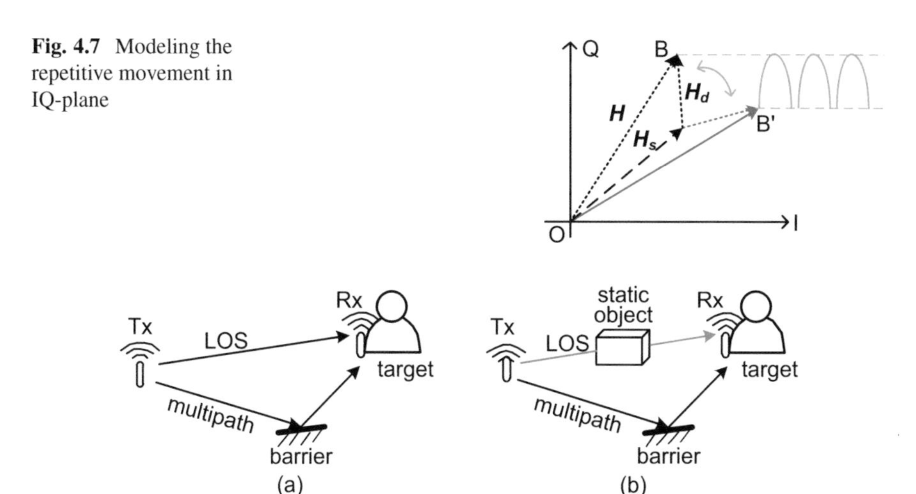

Fig. 4.7 Modeling the repetitive movement in IQ-plane

Fig. 4.8 Effect from surrounding objects for the device-based configuration: (**a**) multipath signal added with the LOS signal, (**b**) LOS signal is weakened by the object

speed components, the spectrogram, which can be obtained via short-time Fourier transformation (STFT), is usually adopted for analysis.

Wireless signals can also be employed to estimate many other targets and environment indicators based on the abovementioned typical physical indicators. For example, the environment temperature can be derived from the propagation speed of the acoustic signal in the air [4]. The dielectric constant of an object can be extracted from the wireless signal channel response [5].

4.2.1.4 Multipath Effect on Sensing the Target

In real sensing environments, there can be many non-target barriers and reflectors that bring noises when sensing the target. When designing a wireless sensing system, we need to carefully consider the environmental effect. To understand the environmental effect, we discuss it in relation to the system configuration, including device-based and device-free configurations.

Under the device-based configuration, the LOS signal between the transmitter and receiver contains the sensing target's information. However, many surrounding objects may exist in the same environment. The surrounding objects in the environment not only weaken the LOS signal but also introduce multipath signals in the received signal. As shown in Fig. 4.8a, the barrier at the bottom brings multipath signals that will be superimposed with the LOS signal at the receiver. Sometimes the multipath signal is stronger than the LOS signal, as depicted in Fig. 4.8b. In this case, the LOS signal is weakened after traveling through the static object.

If we want to measure the propagation time/distance of the LOS signal, the LOS signal and multipath signal should be separated from the received signal. If the

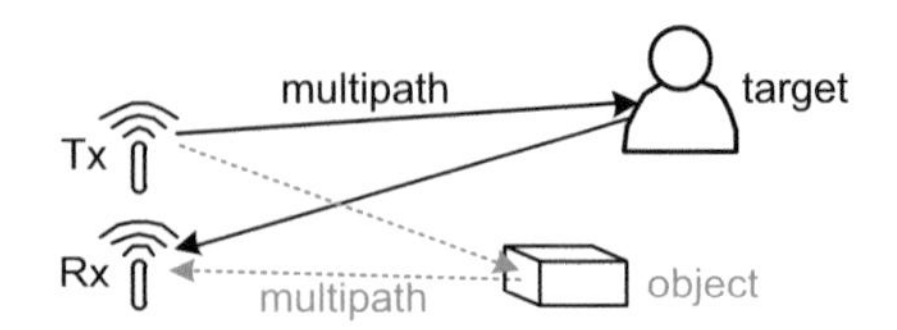

Fig. 4.9 Effect from surrounding objects for the device-free configuration

multipath signal's power is more dominant than the LOS signal, it will be more difficult to separate the LOS signal. This situation may result in the inaccurate detection of the target. Furthermore, the environment layout can change from time to time, making the pattern of multipath signals change frequently. If we simply extract features from the received signal, the effect of multipath signals is also involved. The environment-dependent features will lose their effectiveness once the environment is changed.

The device-free configuration encounters a similar problem as the device-based one. The difference is that the signal from the target is also the multipath signal, as depicted in Fig. 4.9. In addition to the multipath signal reflected by the target, the object beneath the target also introduces multipath signals, which will add more difficulties for separating the target signal from other multipath signals. In the device-based configuration, the LOS signal has the smallest traveling distance in contrast to multipath signals, which can aid the extraction of the LOS signal. This condition does not exist in the device-free configuration. Therefore, the device-free configuration is mainly used for sensing a moving target.

In summary, the multipath effect impedes to achieve accurate sensing of the target. Therefore, we need to remove the effect from multipath signals. For static multipath signals, their effects can be removed by background elimination [6]. When there is no target in the environment, we can measure the background multipath signals in the static environment beforehand, and the acquired measurement can be used to perform background subtraction when the target appears. However, this method is not flexible when the environment layout frequently changes because background signals need to be updated once the environment changes.

Another method for reducing the effort of background elimination is to employ a reference signal to receive a similar multipath effect as the target signal. But the reference signal itself is not affected by the target. Then, the multipath signal can be cancelled out by subtracting the reference signal from the target signal. For example, when sensing human activities with RFID, one reference tag can be attached to one part of the human body that does not have a significant movement of the target activity. Then, another tag is attached to the target body part. The reference tag is usually positioned near the target tag so that they can receive similar multipath signals (including both static and dynamic multipath signals). Then, the multipath effect can be mostly removed from the target signal after subtracting the signal from the reference tag.

4.2.1.5 Position Effect on Sensing the Target

Many studies have shown that human activity patterns captured by wireless signals are also subject to the human position with respect to wireless transceivers in the environment. For the same human activity, its signal pattern can vary when the person is located at different positions and orientations relative to the wireless device [7]. Therefore, we need to extract activity features that are independent of the environment and position.

To achieve this goal, instead of extracting statistical features directly from the wireless signal, we can employ frequency-domain features to reflect the activity movement speed information. For example, STFT is employed to extract the speed information from the wireless signal amplitude for large-scale activities, e.g., walking and running. The rationale for explaining why the frequency of the signal amplitude is correlated with the moving speed is as follows. When the body movement causes its affected signal path to change λ, the signal phase will rotate by 2π, making the overall signal amplitude change in full cycle. Then, the number of cycles n in the amplitude within a time window t, from which the amplitude cycle frequency f_{cyc} is calculated, is related to the moving speed v_m.

$$\because v_m = \frac{d_m}{t} \text{ and } d_m = \frac{n \cdot \lambda}{2}$$

$$\therefore v_m = \frac{n \cdot \lambda}{2t}$$

$$\text{then } f_{cyc} = \frac{n}{t} = \frac{2v_m}{\lambda}$$

However, the cycle frequency can still be affected by the human position because, at different positions, the change of signal path incurred by the same movement can vary. Take Fig. 4.10 as an example. When the person faces to the wireless devices and moves to the right with d_m distance, as the distance between the person and wireless devices d_0 is much larger than d_m, the signal propagation path can decrease around $2d_m$. By contrast, when the person faces to the right and moves upward with d_m, the change of signal propagation path is much smaller than d_m.

Therefore, frequency-domain features also lose their effectiveness. To achieve robust activity recognition under different positions, we can look at the activity itself. The inconsistency of signals under different positions is caused by different relative coordinates between the human body and wireless devices. If we view the

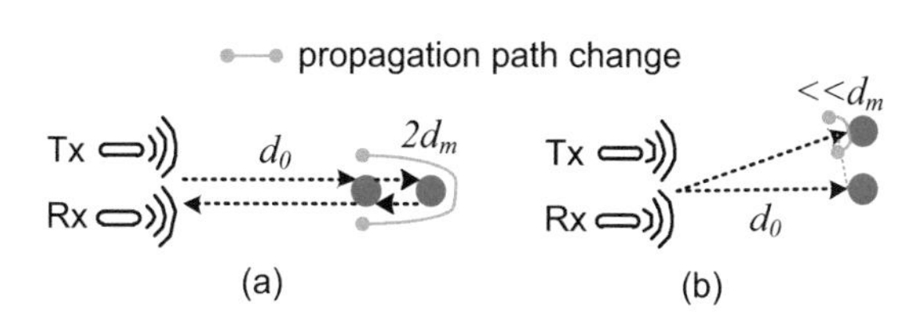

Fig. 4.10 Effect of orientation on the signal propagation path with same moving difference d_m: (**a**) move to the right, (**b**) move upward

activity movement itself, then we can see that its pattern is fixed. Recent studies have attempted to extract activity features from the view of the human body. In gesture recognition, the WiGesture system extracts a position-independent feature, called motion navigation primitive, to capture the direction change of the hand [7]. The Widar3.0 system extracts the body-coordinate velocity profile to characterize the velocity change when performing gestures [8].

4.2.1.6 Sensing Multiple Targets Simultaneously

In previous sections, we mainly introduce how to obtain the information of a single sensing target. In many applications, multiple targets need to be sensed at the same time. Intuitively, multi-target sensing can be achieved via the device-based approach, which requires each target to carry a wireless sensor, e.g., RFID tag or smartphone. Then, each target can be separately sensed using its own sensors. However, device-based sensing is less convenient compared with the device-free way. Thus, it is preferable to adopt the device-free configuration. However, multi-target sensing under the device-free configuration is more complicated. The multipath signals reflected by multiple targets are mixed together in the received signals, as shown in Fig. 4.11. The multipath signals reflected by target 1 ("multipath 1") and target 2 ("multipath 2") are superimposed in the received signal. Each target's signal should be explicitly extracted from the received signal to obtain each target's information. Moreover, the presence of multiple targets, especially among moving targets, mutually affect each other's signal. There may be signals reflected by multiple targets consecutively. As highlighted by the dashed lines in Fig. 4.11, "multipath 3" is jointly composed of the reflection from both target 1 and target 2. The closer are the two targets, the more signals will be simultaneously affected by them. These signals will add more noises to each target's original signal.

To separate multiple targets' signals for the device-free configuration, the straightforward way is to adopt the beamforming technique. For wideband signals, such as FMCW and UWB radars, we can utilize their high spatial resolution to separate multiple targets' signal according to their different distances to the radar. Furthermore, we can employ the multi-antenna design to estimate the target AOA. With the distance and AOA information, the target signal can be explicitly obtained via beamforming. We illustrate the steps of beamforming by using the FMCW radar as an example.

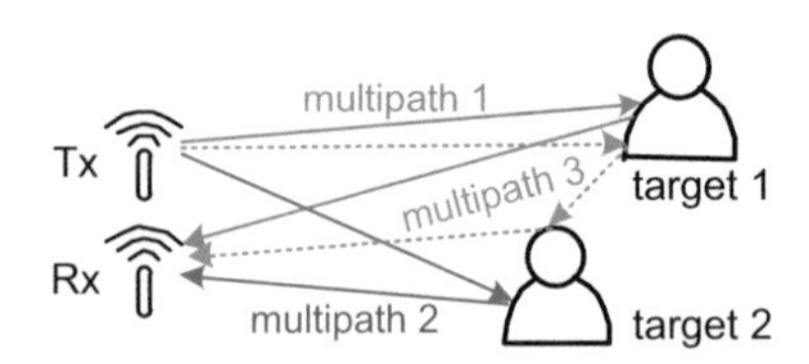

Fig. 4.11 Multi-target sensing (two targets as an example) under the device-free configuration

Assume the radar has M antennas, and the FMCW chirp signals received by M antennas are represented as follows:

$$y_1(t) = A_1 \cdot e^{j2\pi(f_c+Kt)\cdot\frac{d_1(t)}{c}}, \tag{4.4}$$

$$y_2(t) = A_2 \cdot e^{j2\pi(f_c+Kt)\cdot\frac{d_2(t)}{c}}, \tag{4.5}$$

$$\dots \tag{4.6}$$

$$y_M(t) = A_M \cdot e^{j2\pi(f_c+Kt)\cdot\frac{d_M(t)}{c}}. \tag{4.7}$$

The traveling distance from the target to each receiving antenna is given by $d_i(t) = (i-1)\cdot l\cdot cos\theta$, where $i \in (1, M)$, l is the spacing of adjacent antennas, and ϕ is the signal AOA. Then, the phase difference between the ith antenna to the first antenna is expressed as

$$\Delta\varphi_i(\phi) = (i-1)\cdot\frac{2\pi}{\lambda}\cdot d\cdot cos\phi. \tag{4.8}$$

Based on $\Delta\varphi_i(\phi)$, we formulate a steering vector as follows:

$$\omega(\phi) = [e^{j\Delta\varphi_1(\phi)}, e^{j\Delta\varphi_2(\phi)}, \dots, e^{j\Delta\varphi_M(\phi)}]. \tag{4.9}$$

Then, we calculate the summation of the received signal at M antennas and the steering vector as follows:

$$r(\phi, t) = \sum_{i=1}^{M}\omega_i(\phi)\cdot y_i(t). \tag{4.10}$$

Next, we perform FFT on each chirp signal to obtain $R(\phi, f)$. By traversing the ϕ, $R(\phi, f)$ will peak at ϕ_t and the range bin where the target is located. With multiple targets in the environment, there will be multiple corresponding peaks in $R(\phi, f)$. Finally, we can retrieve the signal of target j whose AoA is ϕ_j and range bin is f_j from $R(\phi, f)$ over time.

However, beamforming requires a wide bandwidth and multiple antennas (the number of antennas should exceed four). Thus, this approach is not applicable to narrowband signals and wireless devices with less than four antennas. To tackle the problem of multi-target sensing with narrowband signals, the algorithm called independent component analysis (ICA) is leveraged to separate the signals of different targets [9, 10]. ICA is traditionally used to solve the blind source separation (BSS) problem, which can be defined as follows: there are P independent sources and Q observations of sources S. We can represent Q observations as a $Q \times T$ matrix, where T is the number of timestamps. Then, we define a weight matrix $W_{Q\times P}$, and the observation matrix $R_{Q\times T}$ is calculated as

$$R_{Q\times T} = W_{Q\times P} \cdot S_{P\times T}. \tag{4.11}$$

The objective of the BSS problem is to estimate the matrix W and recover source signals S given the observation R. However, applying the ICA algorithm for signal separation requires source signals to meet two requirements, i.e., the independent and non-Gaussian requirements. Thus, we need to verify or preprocess the wireless signals from multiple targets to comply with the requirements of ICA, which limits the usage of ICA for multi-target sensing.

4.2.2 *Pros and Cons of Model-Based Methodologies*

Establishing models is pivotal to interpret and understand the mechanism and capability of wireless sensing. Based on the signal propagation models and physical laws, we can explain why wireless signals can sense the target and describe how wireless signals are affected by the target. Most importantly, we have shown the effect of multipath signals, target positions, and the surrounding environment on sensing the target. The received signal is the superposition of various factors apart from the target. Thus, we need to explicitly extract the target's information and remove the effect from other factors. In this way, our results can be more robust from other impacts.

However, in practice, many target's behavior and sensing environments are actually too complex to model. Moreover, due to signal imperfections and limits, we may fail to acquire the desired information accurately from wireless signals as modelled. Nevertheless, the information of the target is still buried in the wireless signal. There is a potential to dig out the information using data-driven methodologies.

4.3 Data-Driven Methodologies

The recent advances in data-driven techniques, e.g., data mining and machine learning, can help to extract the target information from noisy signals. The benefit of using data-driven techniques is that, by "seeing" a large number of data samples, they can learn the intrinsic relationship between the data and the results.

The general workflow of the data-driven wireless sensing framework is summarized in Fig. 4.12. The first step is to collect wireless signals and extract signal indicators. Then, signals are pre-processed for noise filtering, segmentation, or other purposes. Next, we extract features from pre-processed signals, which can be achieved by hand-crafting features based on the domain knowledge of the signal pattern or automatically extracted through neural networks. Finally, a learning model is trained for classification, regression, or other tasks. There are mainly two methods of data-based wireless sensing. The first method is to employ traditional

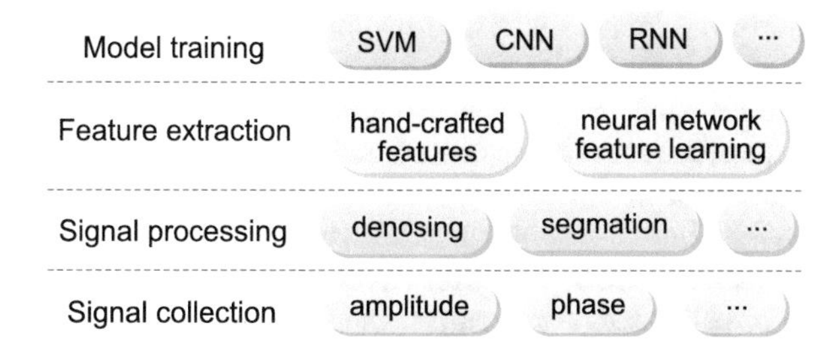

Fig. 4.12 Workflow of data-driven sensing methodologies

machine learning models (e.g., support vector machine (SVM), random forest (RM), and linear regression (LR)), and the other method is to leverage the deep learning models. In this section, we will introduce how learning techniques promote wireless sensing.

4.3.1 Data Analytics and Machine Learning Algorithms for Wireless Sensing

Traditional machine learning algorithms require handcrafted features to train the model. There are mainly four types of features, namely, the time-domain, frequency-domain, time-frequency domain, and spatial domain features. We list the key features of each domain as follows.

- Time-domain features: Many common statistics are employed as time-domain features. For example, the maximum, minimum, mean, variance, kurtosis, skewness, root mean square, percentiles, and zero-crossing rate, are extracted from the temporal signal. The temporal signal can also be directly used to generate a template waveform for each kind of target status or movement.
- Frequency-domain features: For the moving target, its movement pattern can be investigated in the frequency domain, e.g., moving frequency components. A variety of features, such as power spectral density and spectral entropy, are commonly used in the frequency domain. As frequency-domain features are capable of characterizing the intrinsic movement pattern of the target, it is more robust to the environmental changes compared with time-domain features. There are many tricks in obtaining the frequency-domain signal via FFT. First, before performing FFT, we need to subtract the DC component in the signal, which is the mean of the signal. Second, we also add a window, e.g., Hamming window, on the signal to avoid frequency leakage.
- Time-frequency domain features: When the target is moving irregularly, it is more desirable for the extracted features to contain the frequency components at different times. Time-frequency domain features are designed to achieve this

purpose. Short-time frequency transformation and discrete wavelet transformation are typical methods to extract time-frequency domain features, which usually involve two-dimension information. However, there is a tradeoff between the temporal and spectral resolution, which requires careful design based on the application requirement.

- Spatial-temporal domain features: For UWB/FMCW radar or wireless devices with an antenna array, the spatial information of the target, although not precise enough, can be estimated. For example, the signal traveling distance and AOA can be employed as spatial-domain features. In this way, we can acquire spatial information over time, which forms the spatial-temporal features.

As mentioned above, a large number of features can be extracted. However, it is hard to decide what kinds of features offer the best performance for the sensing task. Therefore, feature selection is commonly needed. Feature selection methods consist of unsupervised and supervised methods. In unsupervised methods, the correlation of the feature to the sensing task is compared to remove redundant features. For supervised methods, feature importance can be calculated using algorithms such as the random forest.

After feature extraction and selection, features are input to machine learning models. Popular models for traditional machine learning are k-nearest neighbor (KNN), LR, decision tree, RD, and SVM. These models can also be combined together, which is called ensemble learning, to train an overall model. The benefit of traditional machine learning is twofold. First, it is less computation-intensive so that these models can be easily deployed on resource-constrained mobile devices. Second, they do not require the collection of a huge amount of training data, which saves the training cost.

Traditional machine learning requires careful feature extraction and selection. Much useful information can be lost by crafting heuristic features only based on statistical, spectral, or spatial patterns. The appealing feature of deep learning is that the layer-by-layer neuron structure enables the model to learn features automatically. The deep model has a strong ability to learn descriptive features from complex data. In many sensing tasks with complex target behavior or environments, deep learning has been proven effective in achieving the sensing goal even though wireless signals suffer from noises [11].

In the following, we introduce several popular deep learning models and explain how they are employed in wireless sensing.

- Multi-layer perceptrons (MLP): As the initial version of artificial neural networks, MLP is composed of more than one hidden layer.[3] Figure 4.13a shows a simple MLP model. The power of the hidden layer is to learn what information in the input is useful and how they interact with each other. MLP is mainly used for classification tasks and achieves higher accuracy compared with traditional machine learning algorithms. In wireless sensing systems, MLP is usually

[3] The input vector and output are usually called "visible layer".

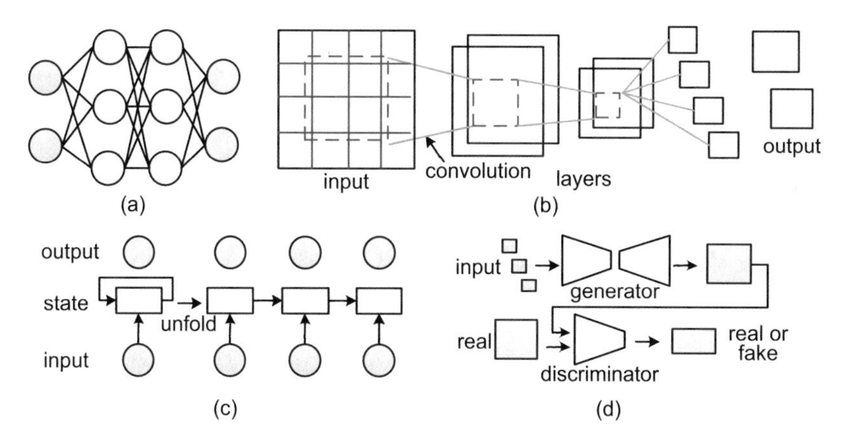

Fig. 4.13 The example architecture of deep learning models: (**a**) MLP, (**b**) CNN, (**c**) RNN, and (**d**) GAN

employed when wireless signals can be formulated into a vector, especially for the vector exhibiting a nonlinear pattern, such as frequency responses of wireless signals among multiple frequencies. By utilizing the nonlinear activation functions, e.g., rectified linear unit (ReLu), MLP can characterize the input vector in a better way. The last layer of MLP can be a softmax layer for classification purposes.

- Convolutional neural network (CNN): CNN has gained much success in computer vision tasks. The key component of CNN is the convolutional kernel, which captures the local relationship between neighbors, as shown in Fig. 4.13b. The convolution operation enables the CNN model to work adaptively to inputs with varying sizes, scales, and translations. In wireless sensing systems, CNN can be applied on time-frequency and spatial domain inputs to extract spatial features. The matrix after performing the short-time Fourier transforming (STFT) is often used as the input of CNN for activity recognition tasks in many wireless sensing applications. Other forms of features, such as the wireless heatmap of two or three-dimension spatial distribution, can be used to recognize the pattern of the target.
- Recurrent neural network (RNN): Different from the CNN that is used more often for spatial inputs, RNN is designed to deal with temporal or sequential data. Figure 4.13c shows the architecture of an example RNN model. The sequence of neurons in RNN enables the memory ability from sequential input. Each neuron's current state relies on its previous states. Meanwhile, the output of each neuron is sent back to the network in a loop. However, RNN suffers from the long-term dependency problem. In other words, we sometimes only need the recent information to predict the next step instead of the whole history. Aiming to tackle this problem, the long-short term memory (LSTM) model was developed. In some sensing tasks, a series of wireless signal indicators, e.g., the signal phase, is input into the RNN model to estimate the AOA of the sensing target.

- Generative adversarial network (GAN): The two models in GAN, are the generative model and the discriminative model, as shown in Fig. 4.13d. The generative model synthesizes a fake sample from the random input. The fake sample is then input to the discriminator, which is trained to differentiate the fake sample from the real one. In wireless sensing, GAN is usually jointly used with other models, such as CNN and RNN, to learn the hidden relationship between the input and output. When it is labor-intensive to collect a huge amount of wireless signal samples for training, GAN can also be utilized to achieve data augmentation. This approach generates synthesized samples following a similar distribution of the actual collected sampling to alleviate the issue of massive data collection.

Deep learning is not limited to the abovementioned models for wireless sensing. The model can be designed in various forms to suit the task challenge and requirement. In the following, we will discuss some typical challenges that can be addressed by data-driven methodologies.

The first challenge is the multi-path effect in different environments. The change of surrounding environments results in different multipath signals that are superimposed with the signal from the target. Thus, the multipath effect should be removed, otherwise the target information can be inconsistent when the environment changes. The key idea of the data-driven methodology to solve this problem is to design a learning model for learning to adapt to new environments. The transfer learning and meta learning techniques can achieve this goal. In transfer learning, we first obtain a pre-trained model (or called the base network) by using data samples from certain environments. Then, in a new environment, we collect a few data samples to retrain or fine-tune the original model. In this way, we do not need to collect a large number of data samples for the new environment. The CrossSense system employs transfer learning to reduce the number of samples needed to retrain the activity recognition model for the new subject and environment [12]. In the Metasense system, the meta learning, also called "learning to learn", enables an adaptive activity recognition by using a few new shots from new environments and subjects [13]. Note that existing data-driven methods mainly attempt to decrease the number of new samples in new environments and subjects. If we want to achieve high-accuracy activity sensing without collecting any new sample, we think model-based solutions may still be required.

The second challenge is the increasing sensing tasks. For human activity sensing, the number of activities to be recognized may increase from time to time. It is cumbersome to train a new model every time a new activity is added. To address this problem, the continual learning technique is adopted. Continual learning, also called lifelong learning, aims to retain old knowledge and accumulate new knowledge by continually learning new tasks over time. When new activities need to be added, we can update the old model only by using data samples collected from the new activities instead of retraining the model by using all kinds of activities' data samples. However, continual learning is mainly used in sensor-based human activity recognition [14]. Its applicability for wireless human sensing has not been widely studied, which calls for further investigation.

The third challenge is the low-resolution signal indicators. The signal indicators estimated by model-based methodologies can be coarse-grained, which may fail to clearly describe the sensing target. We can apply deep learning models to enhance the sensing performance using low-resolution signal indicators. The technique, called cross-modality learning, in which a "teacher–student" network is built to improve the sensing capability of wireless signals. For example, researchers use the images of human skeleton movements as the reference modality data (i.e., "teacher") to guide the training of the learning model to track human movements via wireless signals (i.e., "student") [15].

4.3.2 *Pros and Cons of Data-Driven Methodologies*

Data-driven methodologies, owing to their ability in learning the hidden correlation in the wireless signal, can be utilized for more complex sensing tasks. Data mining and machine learning algorithms can help realize high-performance sensing under changing environments, increasing sensing tasks, concurrent sensing targets. Many feature engineering and data representation methods can help to remove signal noises. Hence, sensing performance can also be guaranteed even when low-cost and commercial wireless devices are used.

However, a large amount of training data need to be collected for data-driven methodologies. In practice, it is difficult to acquire enough data to guarantee the performance of data-driven methodologies. Besides, the model may also learn some features that are irrelevant to the sensing task. For example, for human activity recognition, the model not only learns how to classify different activities but also learns the target-specific feature, making the model ineffective for other people.

As we have introduced model-based and data-driven methodologies for wireless sensing. We know that model-based and data-driven methodologies have their unique benefits but suffer from different problems. Can we take advantage of both model-based and data-driven methods? Many recent studies have given the answer. Researchers have proposed a hybrid way to combine both methods. Currently, there are mainly two ways to combine model-based and data-driven methods. The first way is based on the idea that wireless signals are prone to the effect of noises and irrelevant multipath signals in real environments. As demonstrated by many researchers, the environment and position of the target and wireless devices can easily affect the pattern of the sensing target [8]. If such noises and disturbances are not properly removed, they will have adverse effects on the latter training of learning models. Therefore, we first carefully model the signal propagation for the specific sensing task and propose model-based methodologies to extract critical physical indicators that can exclusively represent the target movements, e.g., moving speed and displacement change. After eliminating as much as noise as possible, the trained model becomes more robust in the face of unpredictable noises and changing environments.

The second way is to add model-based constraints when training the learning model. The essential part of a machine learning model is the loss function, which describes the optimization target of the learning algorithm. If the sensing target's moving pattern or status obeys a particular model-based law, we may convert such a law into a regularizer and add it into the loss function. The most commonly used model-based law is to find a geometric relationship between the sensing target and reference points. By constraining the loss function, the model can be trained to conform to the given law and be less disturbed by other noises and effects.

References

1. Zheng T et al (2020) V2iFi: in-vehicle vital sign monitoring via compact RF sensing. Proc ACM Interactive Mob Wearable Ubiquitous Technol 4(2):1–27
2. Sun K et al (2018) Vskin: Sensing touch gestures on surfaces of mobile devices using acoustic signals. In: Proceedings of the 24th annual international conference on mobile computing and networking, pp 591–605
3. Wang W et al (2015) Understanding and modeling of wifi signal based human activity recognition. In: Proceedings of the 21st annual international conference on mobile computing and networking, pp 65–76
4. Cai C et al (2020) AcuTe: acoustic thermometer empowered by a single smartphone. In: Proceedings of the 18th conference on embedded networked sensor systems, pp 28–41
5. Dhenkne A et al (2018) Liquid: A wireless liquid identifier. In: Proceedings of the 16th annual international conference on mobile systems, applications, and services, pp 442–454
6. Yang L et al (2015) See through walls with COTS RFID system! In: Proceedings of the 21st annual international conference on mobile computing and networking, pp 487–499
7. Gao R et al (2021) Towards position-independent sensing for gesture recognition with Wi-Fi. Proc ACM Interactive Mob Wearable Ubiquitous Technol 5(2):1–28
8. Zheng Y et al (2019) Zero-effort cross-domain gesture recognition with Wi-Fi. In: Proceedings of the 17th annual international conference on mobile systems, applications, and services, pp 313–325
9. Yue S et al (2018) Extracting multi-person respiration from entangled rf signals. Proc ACM Interactive Mob Wearable Ubiquitous Technol 2(2):1–22
10. Zeng Y et al (2020) MultiSense: Enabling multi-person respiration sensing with commodity wifi. Proc ACM Interactive Mob Wearable Ubiquitous Technol 4(3):1–29
11. Li C et al (2021) Deep AI enabled ubiquitous wireless sensing: a survey. ACM Comput Surv 54(2):1–35
12. Zhang J et al (2018) CrossSense: Towards cross-site and large-scale WiFi sensing. In: Proceedings of the 24th annusal international conference on mobile computing and networking, pp 305–320
13. Gong T et al (2019) Metasense: few-shot adaptation to untrained conditions in deep mobile sensing. In: Proceedings of the 17th conference on embedded networked sensor systems, pp 110–123
14. Jha S et al (2021) Continual learning in sensor-based human activity recognition: An empirical benchmark analysis. Inf Sci 575:1–21
15. Zhao M et al (2018) Through-wall human pose estimation using radio signals. In: Proceedings of the IEEE conference on computer vision and pattern recognition, pp 7356–7365

Chapter 5
Case Studies

5.1 Human Respiration Monitoring

As an essential vital sign, respiration is highly related to the overall homeostatic control for human health. Millions of people are suffering from chronic respiratory diseases. Furthermore, respiration is also a critical indicator for other diseases, e.g., cancer and kidney diseases. Therefore, accurate and continuous respiration monitoring (RM) is highly demanded by people to detect and treat diseases timely. We have investigated the usage of the WiFi, RFID, and UWB radar signals for RM under different scenarios, e.g., sleeping, sitting and exercising. We mainly adopt the model-based methodologies to understand how the wireless signal amplitude and phase is affected by respiration activity and derive the periodic respiration pattern. In addition, we also propose model-based solutions to tackle the key challenges in realized respiration monitoring in practical and complex scenarios.

5.1.1 WiFi-Based Respiration Monitoring During Sleep

It is well recognized that sleep affects the productivity or physical vitality of a person and is related to many diseases, including diabetes, depression, or even stroke and heart failure. Therefore, a practical sleep monitoring system is highly desirable. Furthermore, providing accurate respiration information is usually essential for a sleep monitoring system. Therefore, we propose a new RM system leveraging off-the-shelf WiFi devices. This system continuously collects the channel frequency response (CFR) of the WiFi signal and extracts breath information. We can also provide breathing information under different sleeping positions.

J. Cao, Y. Yang, *Wireless Sensing*, Wireless Networks,
https://doi.org/10.1007/978-3-031-08345-7_5

5.1.1.1 CSI for Tracking Respiration

CFR describes how an RF signal propagates from the Tx(s) to the Rx(s) and reveals the combined effect of scattering, fading, and power decay with distance. With commodity WiFi Network Interface Cards (NICs), a group of 30 subcarriers' measurements can be revealed to upper layer users in the format of CFR. Each CFR depicts the amplitude and phase of a subcarrier as

$$H(f_k) = ||H(f_k)||e^{j sin(\angle H_k)}. \tag{5.1}$$

where $H(f_k)$ is the CFR at the subcarrier k with central frequency of f_k, and $||H(f_k)||$ and $\angle H_k$ denote its amplitude and phase, respectively.

From each packet received by Rx, we can extract a 30-by-3 matrix called channel frequency response (CFR) matrix with three receiving antennas. Each column of the CFR matrix corresponds to one antenna while each row corresponds to one subcarrier. Let $CFR^j(t)$ denote the jth column of the matrix extracted from the tth packet received as follows:

$$CFR^j(t) = [h_1^j(t), h_2^j(t), \cdots, h_{30}^j(t)]^T, \tag{5.2}$$

where $h_i^j(t)$ is the CFR on the ith subcarrier at time instant t of antenna j. Note that $h_i^j(t)$ is a complex number and is represented by the amplitude $|h_i^j|$ and the phase δh_i^j as $h_i^j = |h_i^j| * e^{j\delta h_i^j}$.

To analyze the temporal information of $CFR^j(t)$, we put together $CFR^j(t)$ collected at different time, denoted as

$$CFR^j = [CFR^j(1), CFR^j(2), \cdots, CFR^j(m)]. \tag{5.3}$$

Note that CFR^j is a 30-by-m matrix, where m is the number of packets received. Each row of CFR^j represents the temporal change of the CSI information over one subcarrier.

Figure 5.1 shows the amplitude and phase of CFR^1, using data collected in 70 s. From the amplitudes of CFR^1, we can clearly see some ripple-like pattern, which corresponds to the movement of chest. On the other hand, the phase of the CFR^1 seems to be random and does not show clear correlation with breathing. Likewise, Fig. 5.2 shows the amplitudes of CFR^2 and CFR^3. The breathing-caused ripples can also be observed in CFR^3.

To see whether these ripples are caused by the person's chest movement, we select the CFR sequence from the tenth row from CFR^1 (i.e., the time history of the CSI information of antenna 1 at the 10th subcarrier). Figure 5.3 compares the amplitude of a CFR sequence with the acceleration data recorded in the meantime. It can be seen that they exhibit a strong correlation with each other.

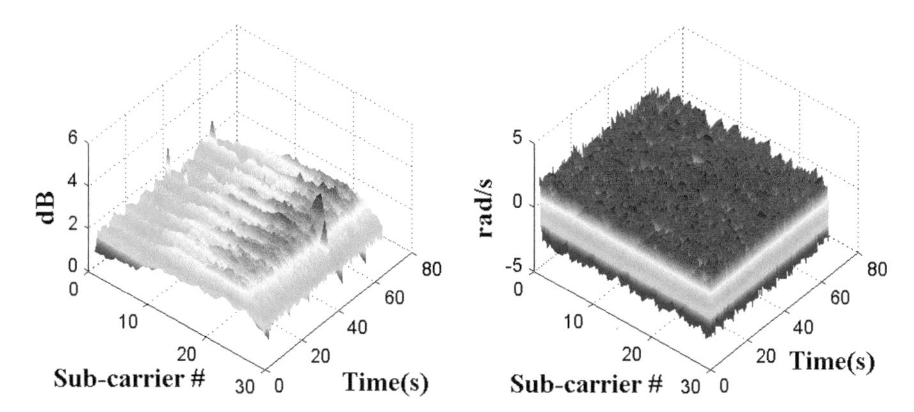

Fig. 5.1 The CFR amplitude and phase obtained during the experiment (antenna 1). (**a**) CFR amplitude (antenna 1). (**b**) CFR phase (antenna 1)

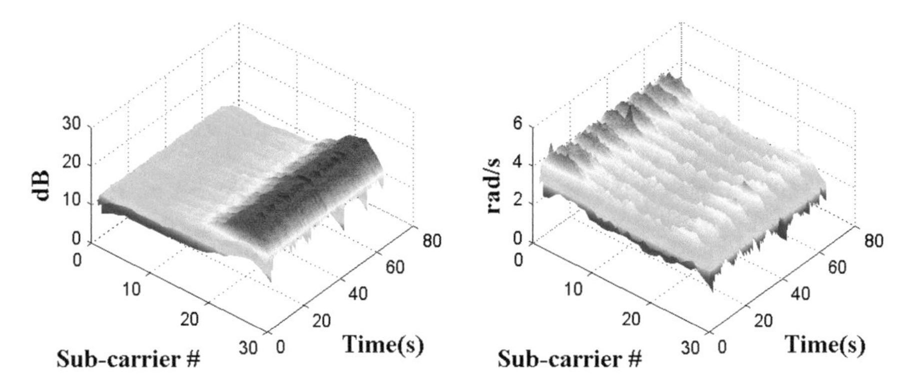

Fig. 5.2 The CFR amplitudes from antenna 2 and 3. (**a**) CFR amplitude (antenna 2). (**b**) CFR amplitude (antenna 3)

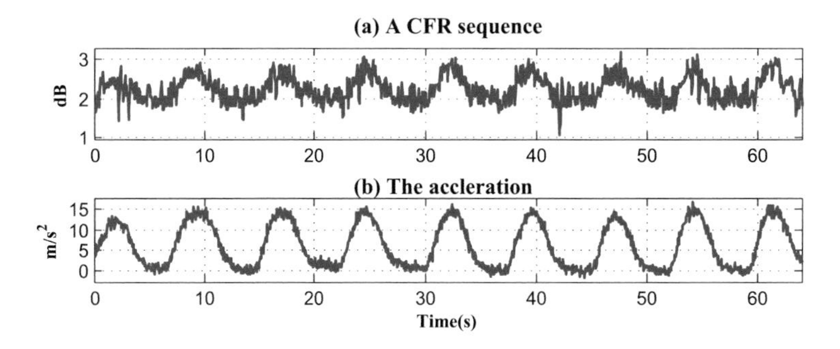

Fig. 5.3 A comparison of the amplitude of (**a**) a CFR sequence and (**b**) the chest movement. We can see that the two sequences closely correlated with each other

5.1.1.2 CFR Data Pre-processing

Based on the findings shown in Fig. 5.1, the amplitude of the CFR data will be taken as the input for monitoring one's respiration. In this section, we describe how the CFR can be processed for tracking respiration better.

The first step of processing a CFR sequence is to remove outliers. We find that in the collected CFR sequences, some abrupt changes in CFR amplitudes are obviously not caused by the movement of chest. Figure 5.4a shows the CFR from all the 30 subcarriers of antenna 1. It can be seen that near 22, 28, and 30 s, there are some significant abrupt changes of the CFR in some or all the subcarriers. We can also see some similar change points in Fig. 5.1. These are outliers for tracking chest movement and must be eliminated.

We utilize the Hampel identifier, which declares any point falling out of the closed interval $[\mu - \gamma * \sigma, \mu + \gamma * \sigma])$ as an outlier, where μ and σ are the median and the median absolute deviation (MAD) of the data sequence, respectively. The reason why we choose median and MAD instead of commonly used mean and standard deviation is because the latter two parameters are extremely sensitive to the presence of outliers in the data. γ is application dependent, and the most widely used value is 3. We apply Hampel identifier on all the 30 subcarriers. Figure 5.4b shows the results after all the identified outliers have been removed. We can see that the Hampel identifier performs very well in this scenario.

It should be noted that in our previous experiments, although the transmitter is programmed to transmit packets every 50 ms, we cannot guarantee the receiver is able to get one packet with the same frequency. We find that sampling jitter is quite common and sometimes can reach more than 300 ms. Jitters are caused by

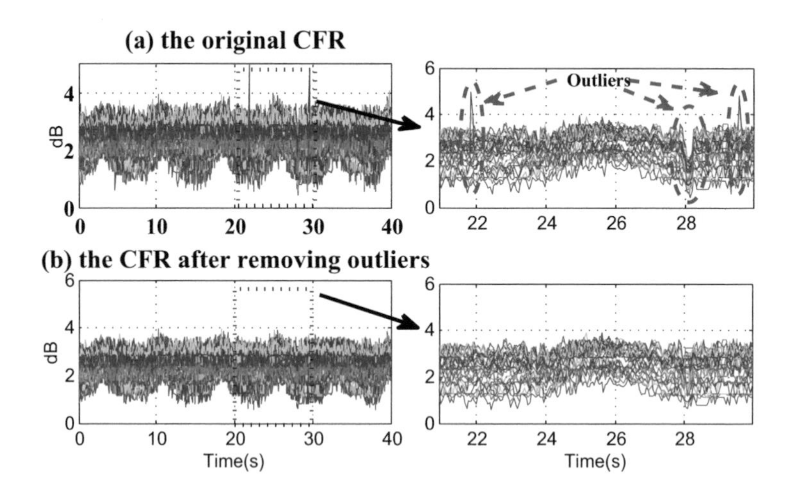

Fig. 5.4 (**a**) The original CFR from all the 30 subcarriers of antenna 1. Left: all the time span (from 0–40 s), Right: selected time span(from 20–30 s). (**b**) The CFR after the outliers are removed using the Hampel filter

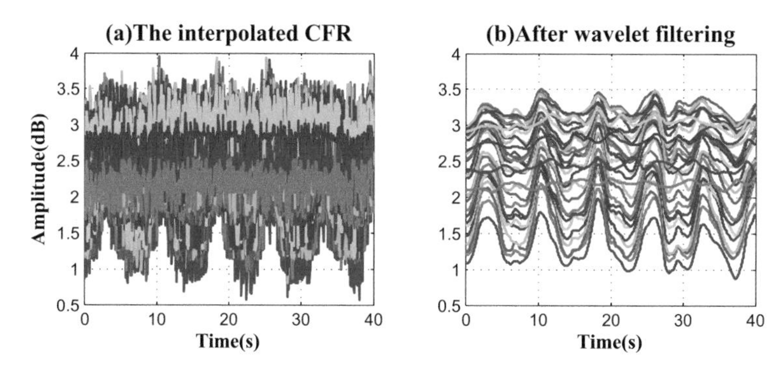

Fig. 5.5 The interpolated 30 CFR sequences (**a**) before using the wavelet filter and (**b**) after using the wavelet filter

packet loss or due to the Linux system on the APs, which lacks support for the users to prioritize high-level tasks arbitrarily. Therefore, the CFR signal must be interpolated. We utilize linear interpolation to obtain the CFR samples at a universe sequence of evenly spaced time points, with 50 ms apart between consecutive values.

After interpolation, the noise contained in the CFR data should be eliminated. We argue that it is not appropriate to use conventional filters (e.g., the Butterworth and Chebyshev filters) to remove high-frequency noise contained in the CFR. This is because they not only smooth away noise but also blur the rising/falling edges that possibly appeared in CFR signals, which is, however, critical for detecting sleep apnea and rollovers. Here, we apply the wavelet filter since it can preserve extremely well the sharp transitions in signals than the other low-pass filters. To be more specific, we apply a 4-level "db4" wavelet transform on each CFR sequence and use only the approximation coefficients to "re-construct" the filtered signal. As an illustration, Fig. 5.5a, b show all the 30 CFR sequences before and after using the wavelet filtering. It can be clearly seen that the original noisy CFR signals become much cleaner.

5.1.1.3 Breathing Rate Estimation on Individual CFR Sequence

We can obtain the respiration rate for each CFR sequence by performing a short-time Fourier transform (STFT). The idea of the STFT is to divide a time signal into shorter segments and then compute the fast Fourier transform (FFT) on each segment. The STFT can reveal how the frequency content of the signal changes with time and therefore is adopted for tracking respiration. Intuitively, the breathing rate can be identified as the peak of the FFT. For example, Fig. 5.6 shows the FFT of a CFR sequence over a 40-s window, and the identified breathing rate is 0.125 Hz.

However, using the above approach usually suffers from a low frequency resolution. For example, the CFR sequence in Fig. 5.6 contains data of 40 s, which

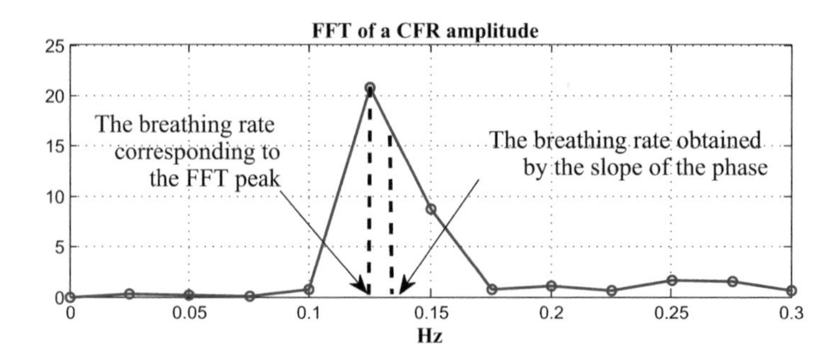

Fig. 5.6 The FFT of the amplitude of a CFR sequence. The location of the peak is the breathing rate (0.125 Hz). This is however a coarse estimation of the breathing rate considering the frequency resolution is only 0.025 Hz. However, using the method proposed in [22], a more accurate respiration rate (0.133 Hz) is obtained

leads to the 0.025 Hz frequency resolution. Larger time duration can provide higher frequency resolution but will decrease time resolution, making the system less capable of tracking changes in breathing rate. To handle this problem, instead of using the FFT peak, we adopt the method proposed in [22] and [1] to estimate the breathing rate. We first implement the FFT for each segment but keep only the peak and its two adjacent bins. Then, we perform an inverse FFT to obtain a complex time-domain signal. The phase of the signal will be linear, and its slope will correspond to the breathing rate. Using the approach above can achieve respiration rate estimation with higher frequency resolution. For example, Fig. 5.6 shows that using the above method, the respiration rate is 0.0133 Hz, which represents a higher frequency resolution and is more accurate than the FFT method.

Since we have multiple subcarriers, we will show how to combine the information from different subcarriers to obtain a better result. Due to frequency-selective fading, the breathing motion will cause different extents of signal perturbation on different subcarriers. As can be seen from Fig. 5.5b, not all the CFR sequences show the pattern of breathing equally well. For example, Fig. 5.7a shows 5 CFR sequences from subcarriers #1, #5, #15, #25, #30. We can see that some of CFR sequences, particularly for subcarrier #30 (signal shown in the middle of Fig. 5.7a), do not contain as much information of breathing as others. Considering that the breathing motion may have little effect on a specific subcarrier, utilizing the information from multiple subcarriers would potentially improve the accuracy as well as the reliability of breathing rate estimation.

For the CFR sequence of each subcarrier, we can obtain a respiration rate. For the 30 subcarriers, we therefore have $\mathbf{f}_e = \{f_1, ... f_{30}\}$, with f_i corresponding to the estimate from the ith subcarrier. Ideally, all elements in $\mathbf{f}_e$ should be the same. However, due to the noise caused by selective fading, the breathing rate could be different among different subcarriers.

To take the advantage of multiple subcarriers, we first remove the outliers in $\mathbf{f}_e$ using the modified Z-score test. The modified Z-score of each f_i is defined as

$$Z(f_i) = \frac{0.7645(f_i - median(\mathbf{f}_e))}{median(|\mathbf{f}_e - median(\mathbf{f}_e)|)}. \tag{5.4}$$

Previous studies recommended that the modified Z-scores with an absolute value greater than 3.5 be labeled as outliers.

For the remaining elements after the Z-score test, the weighted median is utilized as the final estimate of breathing rate. Assume we have n distinct elements $f_1, f_2, ..., f_n$ with positive weights $w_1, w_2, ..., w_n$ such that $\sum_{i=1}^{n} w_i = 1$, the weighted median is the element f_k satisfying: $\sum_{f_i<f_k} w_i < 1/2$ and $\sum_{f_i>f_k} w_i \leq 1/2$. The reason why we choose the weighted median here is that it allows for non-uniform statistical weights related to varying precision measurements in the sample. Considering CFR sequences for different subcarriers are affected differently by the respiration, they should be weighted differently for final respiration rate estimation.

Next, we will show how to determine the weight w_i for each CFR sequence. To quantify the periodicity of a CFR sequence, we first model the sequence as a sinusoidal wave. Then the two parameters, the goodness of fit of the model and the amplitude of the sinusoid, are utilized to calculate the periodicity level.

When a person is breathing, a measured CFR sequence $x(t)$ can be largely modeled as a sinusoidal wave.

$$x(t) = Asin(2\pi f t + \phi) + D + \epsilon(t) \tag{5.5}$$

where the constants A, f, ϕ and D are the amplitude, frequency, phase, and shift of the identified sinusoidal wave, and $\epsilon(t)$ is an additive noise.

Given $x(t)$, the parameters of the sinusoidal wave (A, f, ϕ and D) can be identified using the Nelder-Mead method, which is a common non-linear optimization technique for multidimensional unconstrained minimization. Here, considering the frequency f has already been estimated before, we regard f as a known constant when applying the Nelder-Mead method.

Once we have identified the parameters A, ϕ and D, the periodicity of $x(t)$ can be characterized by two factors: the goodness of fit of the sinusoidal wave and the amplitude A. The justification is as follows. First, if a CFR sequence $x(t)$ can be accurately modelled as a sinusoidal wave, then we regard $x(t)$ with a high periodicity level. The goodness of fit can be calculated by the root-mean-square error (RMSE) defined as

$$RMSE = \sqrt{\frac{\sum_{t=1}^{n}(\hat{x}(t) - x(t))^2}{n}}, \tag{5.6}$$

where $\hat{x}(t)$ is predicted values at time t using the sinusoidal model and n is the length of the $x(t)$.

Second, we argue that the larger the amplitude of the identified sinusoidal wave, the higher the periodicity level. Based on the $RMSE$ and A, we define the periodicity level of a $x(t)$, denoted as p_r, as the ratio of the two parameters.

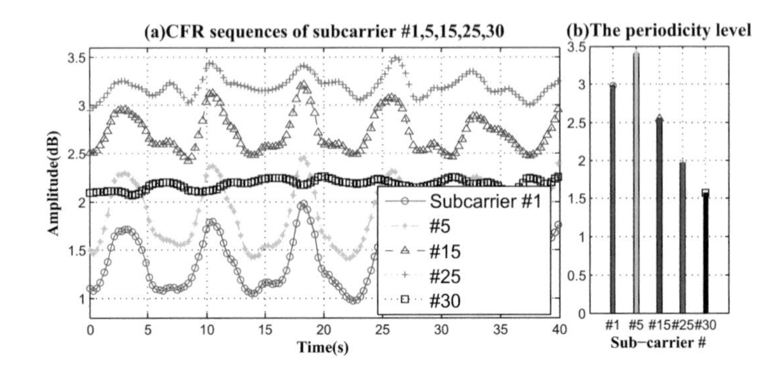

Fig. 5.7 (**a**) The CFR sequences from subcarrier #1, #5, #15, #25, #30, (**b**) The corresponding periodicity level of the five CFR sequences

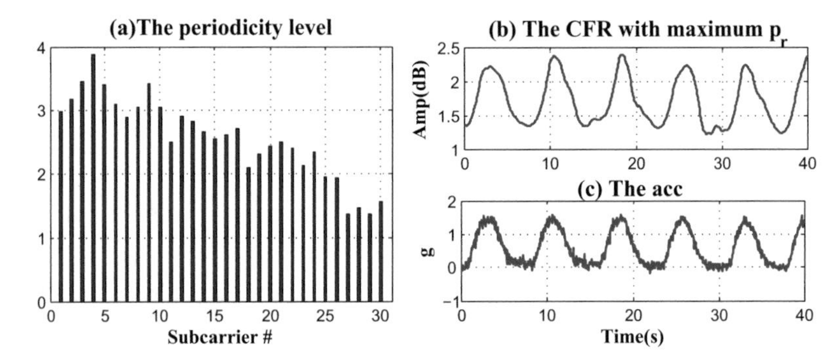

Fig. 5.8 (**a**) The periodicity level of 30 CFR sequences, (**b**) The CFR subcarrier with maximum p_r (i.e., Subcarrier #4), (**c**) The corresponding acceleration data

$$p_r = \frac{A}{RMSE} \tag{5.7}$$

Obviously, the greater the p_r, the higher the periodicity of a CFR sequence.

As an illustration, Fig. 5.7b depicts the value of p_r of the five CFR sequences shown in Fig. 5.7a. It can be seen that the CFR sequences from the first three subcarriers (i.e. #1, #5, #15) have greater p_r than the remaining two. This matches well with the observation that the CFR sequences from the these three subcarriers look more periodic than others.

Figure 5.8a shows the periodicity level of all the 30 CFR sequences shown in Fig. 5.5b. Figure 5.8b shows the CFR from the subcarrier #4, which has the maximum p_r. Figure 5.8c shows the corresponding acceleration data. It can be seen that the CFR from this subcarrier shows high correlation with the breathing-induced motion. This observation justifies that the obtained respiration rate from the subcarriers with higher periodicity level should be assigned with higher weights for respiration monitoring.

After obtaining p_{ri} of a subcarrier i, the w_i for the frequency estimate of this subcarrier is calculated as

$$w_i = \frac{p_{ri}}{\sum_{i=1}^{n} p_{ri}}. \tag{5.8}$$

Please be noted that p_r is utilized not only for estimating the weight of f_i for breathing rate estimation, it is also taken as a feature for identifying the sleeping postures and sleep apnea. This part will be described in the following section.

It is interesting to see what affects the periodicity of each subcarrier. After a number of experiments, we find that if we fix the placement of Tx-Rx pairs and let the sleeper stay in a certain position, the periodicity of each subcarrier generally remains stable with time. However, when the location of Tx-Rx pair or the location of sleeper changes, the periodicity of subcarriers will change accordingly. After analyzing, we find that this is because the change of location of Tx-Rx/sleeper will cause the change in certain multi-path components, which as a consequence, will lead to the change of the frequency-selective fading. As a result, the periodicity of subcarriers will change accordingly.

5.1.1.4 Tracking Respiration at Different Sleeping Positions

In this section, we will study the effect of different sleeping positions. For simplicity, we assume the person breathes normally in these sleeping positions. Figure 5.9 shows 6 most common sleeping positions: (1) Foetus: Sleeping all curled up into a ball with knees drawn up and chin tilted down. (2) Log: On the side, arms at sides. (3) Yearner: On the side, arms out. (4) Soldier: On the back, arms at sides. (5) Freefaller: Face down. (6) Starfish: On the back, arms up.

We test the system when the person is at different sleeping positions. In particular, the Tx and the Rx are placed at two sides of the person and the person is breathing normally for about 10 min. As an example, Fig. 5.10a shows the processed CFR sequences of all the 30 subcarriers in a 30 s period when the person is in

Fig. 5.9 The most popular sleeping positions. From left to right: Foetus, Log, Yearner, Solider, Freefaller, and Starfish. (The figure is from [3])

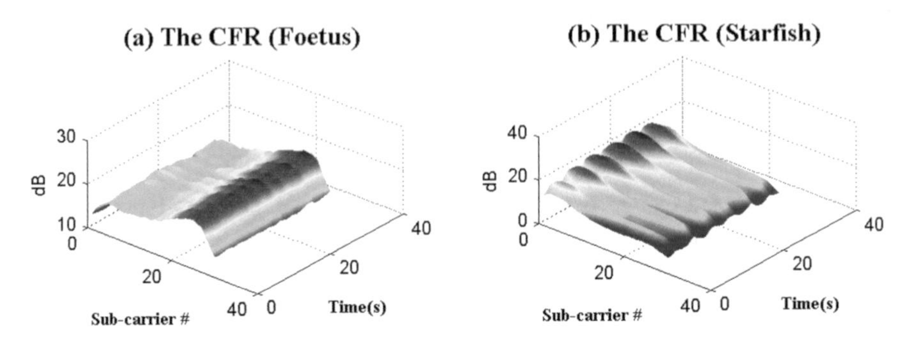

Fig. 5.10 The results of the system for sleeping positions (**a**) "Foetus" and (**b**) "Starfish"

the "Foetus" position. Similarly, Fig. 5.10b shows the corresponding results of the "Starfish" position. Obviously, respiration can be much better tracked when the person is at the "Starfish" position.

What causes the problem? We believe it is caused by the 'mis-match' between the location of the Tx-Rx pair and the person's sleeping position. In these experiments, the Tx and the Rx are placed at different sides of the human body. As shown in Fig. 5.11a, for the sleeping positions like the "Soldier" and "Starfish" in which the person is facing upward, there are many NLOS paths that carry information of the chest movement. These *"effective paths"* are generated when the wireless signal sent from the Tx is reflected from the chest. In this condition, respiration can be well captured using the CSI information. However, for the "Foetus", "Log", and "Yearner" in which the person is sleeping on one side (see Fig. 5.11b), the number of effective paths is much smaller, and the performance will be degraded.

To better track the chest movement under "Foetus", "Log", and "Yearner", we need to change the position of the Tx-Rx pair such that the number of effective paths can be increased. This can be achieved by placing both the Tx and the Rx at the same side of the person's chest, as shown in Fig. 5.11c. As an illustration, Fig. 5.12 shows the results after the Tx-Rx pair is put at the same side of the person's chest. Specifically, Fig. 5.12a, b, and c show the CFR sequence with the maximum periodicity for the sleeping positions "Foetus", "Log", and "Yearner", respectively. It can be seen that with change of the location of the Tx-Rx pair, the performance of the system is significantly improved.

Therefore, we argue that to reliably track a person's breathing, it is preferable to have three Tx-Rx pairs. As shown in Fig. 5.13, the first pair, consisting of transmitter 1 (Tx_1) and the receiver 1 (Rx_1), is placed at the left side of a person to track his respiration when he sleeps on his left side. The second pair, Tx_1-Rx_2, is utilized when he sleeps on his back. Note that these two pairs share the same Tx_1. This can be achieved by letting Tx_1 broadcast WiFi signals that can be received by both Rx_1 and Rx_2. The third pair Tx_2-Rx_3, is placed at the right side of the person to track the respiration when he is sleeping at the right side. It should be noted that we cannot eliminate Rx_3 by letting Rx_2 alternatively receive WiFi signals from Tx_1 and Tx_2 because the handoff usually takes a few seconds.

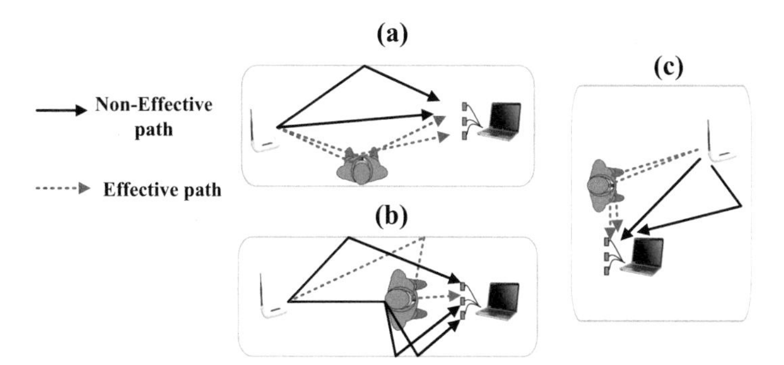

Fig. 5.11 (**a**) For sleeping positions like the "Soldier" and "Starfish", there are many effective paths which carry information of chest movement. (**b**) For sleeping positions like "Foetus", "Log", and "Yearner", the number of "effective paths" is much smaller. (**c**) By changing the positions of the Tx-Rx pair, the number of effective paths can be increased

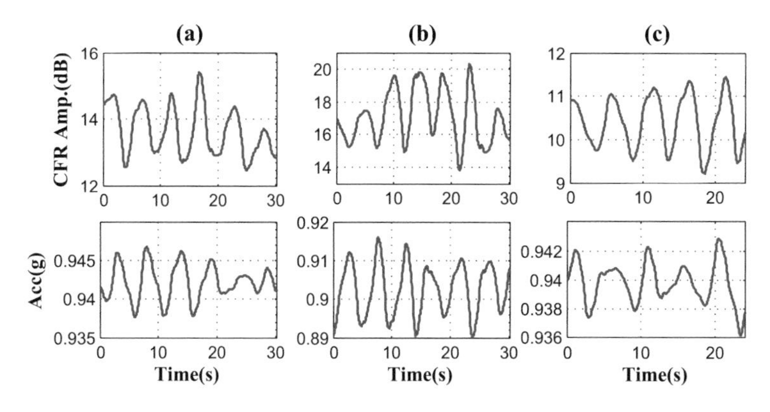

Fig. 5.12 The results of the system for sleeping positions "Foetus","Log", and "Yearner"' after Tx-Rx pair is put at the same side of the person's chest. (**a**) The CFR amplitude from the CFR sequence with the maximum p_r (upper figure) and the corresponding acceleration (lower figure) for the "Foetus" position. (**b**) The CFR and the acceleration for the "Log" position and (**c**) for the "Yearner" position

Fig. 5.13 Three pairs of Tx-Rx collaboratively work to monitor the respiration of a person. Tx_1 and Rx_1 work when the person is sleeping on his left side. Tx_1-Rx_2 work when the person is sleeping on his back, and Tx_2-Rx_3 are utilized for the right-side sleeping positions

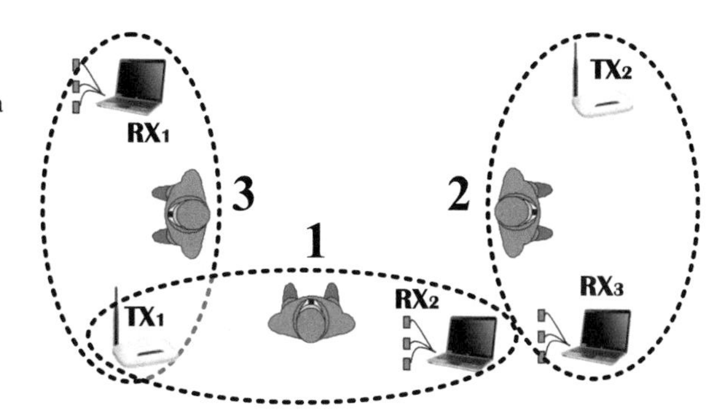

5.1.1.5 Experiments and Evaluation

We prototype our system with commodity WiFi devices and evaluate its performance in a typical office. Two commercial TP-LINK WR740 wireless routers, denoted as Tx_1 and Tx_2, are employed as the transmitters operating in IEEE 802.11n AP mode at 2.4 GHz. Two Lenovo desktop computers, denoted as Rx_1 and Rx_2, are used as receivers, which is equipped with off-the-shelf Intel 5300 network interface card. The transmitter Tx_1 is programmed to broadcast packages to Rx_1 and Rx_2 every 50 ms, and Tx_2 is programmed to send packages to Rx_3 with the same rate. The data is processed by the corresponding laptop. In the meantime, the ground-truth respiration is obtained by a smartphone attached to the person's chest.

We define the metric of accuracy of respiration rate estimation RR as follows. Let the respiration rate estimated from a segment of CFR sequences be f_{CFR}, and the 'true' respiration rate estimated by the smartphone accelerometers be f_{ACC}, then we define the respiration rate is correctly identified if $f_{CFR} \in [0.9 f_{ACC}, 1.1 f_{ACC}]$. For a test in which we have N segments of CFR data, we define $RR = m/N$ as the final metric for respiration estimation, where m is the number of the segments for which the respiration rate is successfully identified.

We first test the system's performance when the person is at fixed six sleeping positions shown in Fig. 5.9. The test lasts about 10 min for each position, and the person is taking normal breathing. For convenience, we let the person always sleep at his left side for positions including "Foetus", "Log", and "Yearner". Therefore, only data from Tx_1-Rx_1 and Tx_1-Rx_2 is utilized in this experiment.

First, we set the window size for respiration estimation as 20 s. Therefore, the CFR sequences in the test are divided into 30 segments. Figure 5.14a, b show the RR for the six sleeping positions using data from Tx_1-Rx_2 and Tx_1-Rx_1, respectively.

First, we can observe that our system can reliably track the respiration rate for different sleeping positions. Both Tx_1-Rx_2 and Tx_1-Rx_1 can give high estimation rate (greater than 85%) for all six sleeping positions. The high accuracy can be

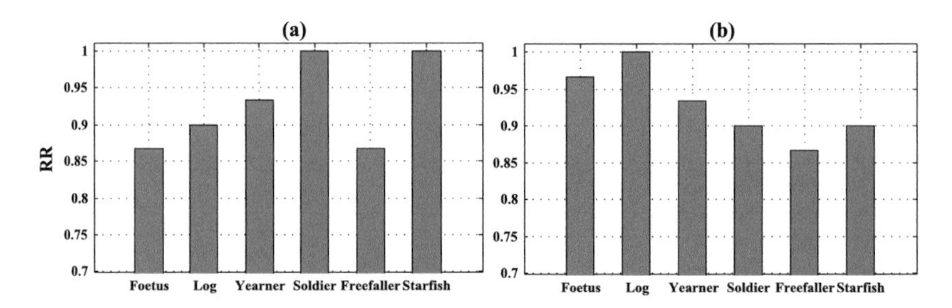

Fig. 5.14 The accuracy of respiration rate estimation under different sleeping positions using data from (**a**) Tx_1-Rx_2 and (**b**) Tx_1-Rx_1. The results show that using data from Tx_1-Rx_2, the respiration can tracked very well when the person is sleeping on his back (i.e., "Soldier" and "Starfish"), while using data from Tx_1-Rx_1 can work well when the person is sleeping on one side (i.e. "Foetus","Log", and "Yearner")

partially attributed to the fact that information from all 30 subcarriers has been efficiently utilized.

Second, it can be seen that when the person is sleeping on his back (e.g. "Soldier" and "Starfish"), Tx_1-Rx_2 performs very well. The respiration rate for all 30 segments are correctly identified, and hence the $RR = 1$. When the person is sleeping on his left side, the RR of using Tx_1-Rx_2 is a little bit lower. On the other hand, Fig. 5.14b shows the results using data from Tx_1-Rx_1. We can see that using data from Tx_1-Rx_1, the person's respiration can be accurately monitored for "Foetus", "Log", and "Yearner". Therefore, if we can accurately identify the sleeping positions, it is expected that one's respiration can be accurately monitored.

5.1.2 RFID-Based Respiration Monitoring in Dynamic Environments

RFID signal has been well investigated for RM due to the small, lightweight, and flexible properties of passive RFID tags, which offer a non-intrusive way for RM by simply attaching RFID tags on the chest [32, 38, 41]. Meanwhile, RFID tags are cost-effective (0.1–0.2 USD per tag) and can be applied for large-scale deployment. The intuition of RFID-based RM is that the tiny periodic chest movement during breathing can be captured by tracking the movement of the tag on the chest. However, current RFID-based RM systems can only monitor the person in a relatively static environment where no people move around so that the tiny chest movement caused by human respiration can be correctly measured. As depicted in Fig. 5.15a, a person is monitored in a static environment, and the measured respiration signal from the RFID tag shows a clear periodic respiration pattern. However, in practice, the environment can be dynamic with people moving nearby. The surrounding moving people could bring multipath signals in the overall received signal from the RFID tag, and the respiration pattern in the RFID signal would be distorted. As shown in Fig. 5.15b, the measured respiration signal becomes noisy.

In this case study, we aim to remove the effect of multipath signals in dynamic environments for realizing robust RFID-based respiration monitoring with accurate apnea detection and respiration rate estimation. To achieve this goal, there are three tasks:

- The first task is to elaborate how the dynamic multipath signals from moving people affect the respiration signal of the monitored person.
- The second task is to remove the effect of multipath signals for accurate apnea detection.
- The third task is to remove the effect of multipath signals for accurate respiration rate estimation.

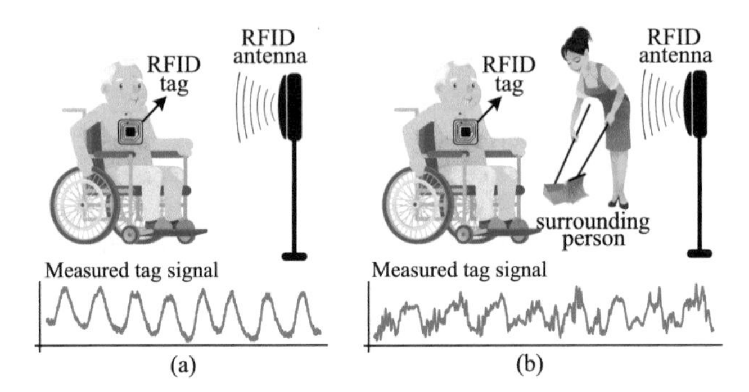

Fig. 5.15 Respiration monitoring in the static and dynamic environments. The signal in the dynamic environment suffers from many noises compared with that in the static environment. (**a**) Static environment. (**b**) Dynamic environment

5.1.2.1 Understanding RFID-Based Respiration Monitoring and the Multipath Effect

To interrogate an RFID tag, the RFID reader first sends out a continuous wave (CW) to activate the tag. After powered up, the tag modulates its information on the CW and reflects it back to the reader. The commodity RFID reader can then extract and output the low-level data of the RFID signal. In our work, we use the RFID signal phase to measure the respiration state, since the signal phase is more sensitive to the minute chest movement during breathing.

To thoroughly understand the RFID signal phase, we interpret it from the aspects of both signal voltage and signal traveling distance. First, we refer to the phasor space, as shown in Fig. 5.16a, to show how the signal phase is measured from the signal voltage. When the RFID reader receives the tag backscattered signal, it is converted into the baseband signal **V**, which can be represented as follows [21]:

$$\mathbf{V} = \mathbf{V}_o + \mathbf{V}_t^i;\ \mathbf{V}_o = \mathbf{V}_{leak} + \mathbf{V}_{scatter}. \tag{5.9}$$

$\mathbf{V}_o$ is decided by the reader transmitter to receiver leakage $\mathbf{V}_{leak}$ and scattering $\mathbf{V}_{scatter}$ from the environment. $\mathbf{V}_t^i$ is the voltage of the tag backscattered signal. $\mathbf{V}_t^i$ changes with the state of the tag chip (i = state 0 or 1). State 1 and state 0 refer to the matching and mismatching states between the input impedance of the tag antenna and the tag chip, respectively. After removing the DC component in **V**, the signal phase φ is calculated as follows:

$$\varphi = ang(\mathbf{V}_t^1 - \mathbf{V}_t^0) = arctan(\frac{Q_{ac}}{I_{ac}}), \tag{5.10}$$

where Q_{ac} and I_{ac} refer to the AC quadrature and in-phase components, respectively. When the tag moves along with the chest movement while breathing, $\mathbf{V}_t^i$ will rotate back and forth, resulting in a periodic change of the signal phase.

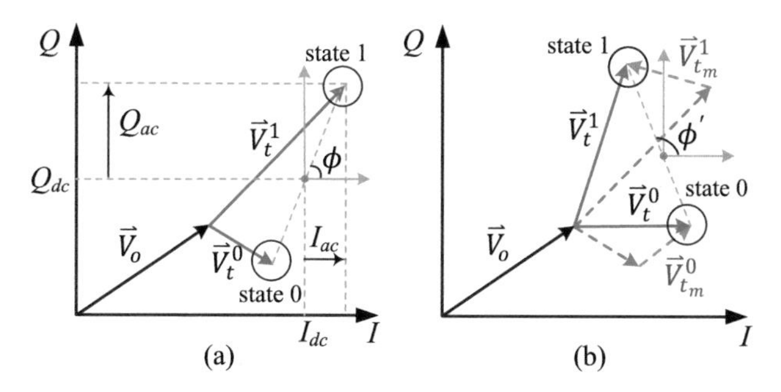

Fig. 5.16 Demodulated voltage of the tag signal received by RFID reader. (**a**) Static state. (**b**) Dynamic state

Second, the signal phase can also be expressed as a function of the signal traveling distance d as follows:

$$\varphi = \{2\pi \cdot \frac{d}{\lambda}\} \mod 2\pi, \tag{5.11}$$

where λ is the signal wavelength. During respiration, with the RFID tag attached on the chest and facing to the antenna directly, Eq. (5.11) becomes

$$\varphi = \{2\pi \cdot \frac{2[d_0 + d_r(t)]}{\lambda}\} \mod 2\pi, \tag{5.12}$$

where d_0 is the initial distance between the tag and the antenna. $d_r(t)$ is a sinusoidal function which describes the chest movement. As the chest moves forward and backward periodically, the signal phase exhibits a periodic pattern accordingly, with valleys and peaks indicating the expansion and contraction of the chest, respectively.

In RFID-based RM systems, the LOS signal between the tag and antenna is used for extracting the respiration pattern [34, 38, 41]. While, in practice, many reflectors in the environment, e.g., surrounding people and furniture, could bring multipath signals. As shown in Fig. 5.17, the static object and moving person bring different multipath signals. Such multipath signals can be superimposed with the LOS signal at the receiver, which greatly affects the signal phase. When surrounding people move nearby the tag, the multipath signals' voltage is added on the received signal, and the tag signal phase φ can be expressed as

$$\varphi = ang[\underbrace{\sum_{s=1}^{S}(\mathbf{V}_{t_s}^1 - \mathbf{V}_{t_s}^0)}_{\mathbf{V}_S} + \underbrace{\sum_{m=1}^{M}(\mathbf{V}_{t_m}^1 - \mathbf{V}_{t_m}^0)}_{\mathbf{V}_M}], \tag{5.13}$$

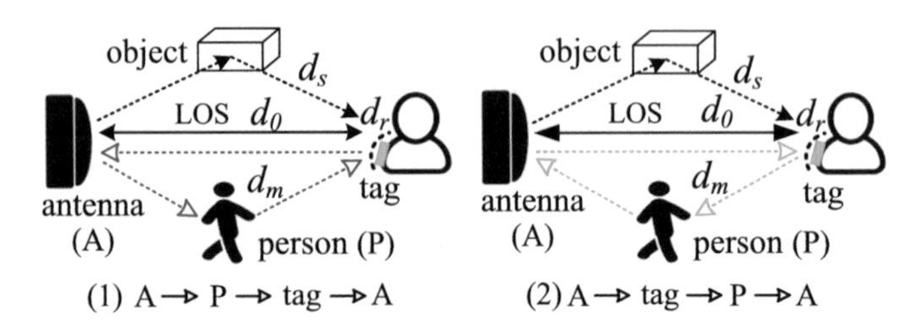

Fig. 5.17 Propagation path of multipath signals from moving person

Fig. 5.18 Effect of dynamic multipath signals on the signal phase

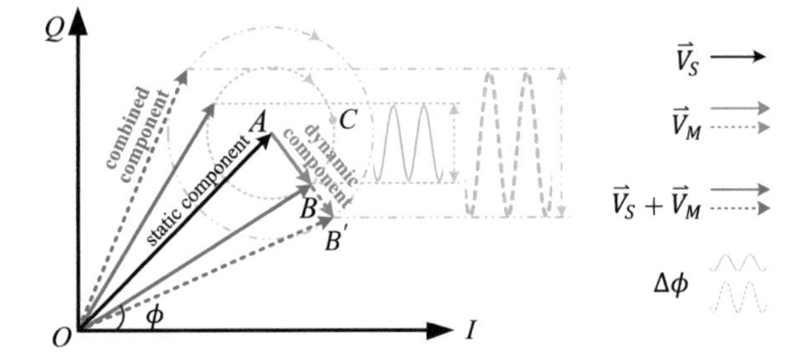

where $\mathbf{V}_{t_s}^{1,0}$ is the voltage of static components, including the static LOS signal and static multipath signals. $\mathbf{V}_{t_m}^{1,0}$ refers to the voltage of dynamic multipath signals. S and M are the total numbers of static signals and dynamic multipath signals in the environment, respectively.

In the IQ diagram, suppose the tag is attached on a static object, **OA** in Fig. 5.18 represents the sum of static components, i.e., $\mathbf{V}_S$ in Eq. (5.13). When people move around the tag, the dynamic component $\mathbf{V}_M$, i.e., **AB** in Fig. 5.18, will rotate from 0 to 2π. The measured signal is denoted by the combined component **OB**. In consequence, the combined signal phase is jointly affected by $|\mathbf{V}_M|$ and $\angle\mathbf{V}_M$ (the angle between $\mathbf{V}_M$ and I-axis). When people move nearby the tag, the strength (length) of $|\mathbf{V}_M|$ varies, e.g., $|\mathbf{V}_M|$ increases from **AB** to **AB′**. Meanwhile, $\angle\mathbf{V}_M$ may also change accordingly, e.g., $\angle\mathbf{V}_M$ decreases when **AB** rotates to **AC**. Then the combined signal phase will change accordingly. Therefore, in the following two subsections, we will study how surrounding people's movements affect the signal phase from the views of $|\mathbf{V}_M|$ and $\angle\mathbf{V}_M$.

Effect of $|\mathbf{V}_M|$

By investigating the effect of surrounding people's movements on $|\mathbf{V}_M|$, we can compare the magnitude of phase changes incurred by respiration with those from people's moving. To achieve this, we look into the propagation paths of multipath signals. As shown in Fig. 5.17, multipath signals primarily propagate in two ways [31]: (1) antenna → person → tag → antenna; (2) antenna → tag → person → antenna. In propagation way (1), the moving person affects the downlink of multipath signals, i.e., [antenna → person → tag]. At this point, $|\mathbf{V}_M|$ mainly depends on the strength of multipath signals, which are dominated by the reflection

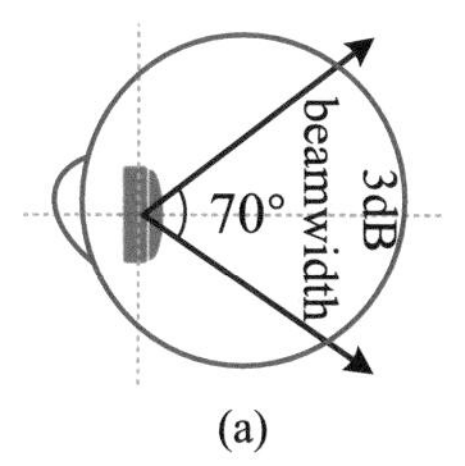

(a)

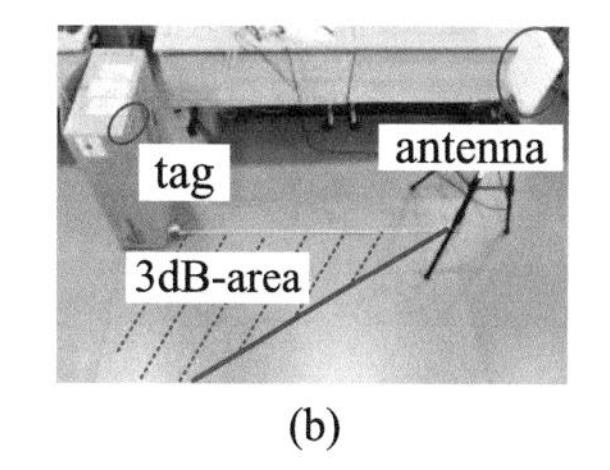

(b)

Fig. 5.19 Illustration of RFID antenna radiation range and 3 dB beamwith. (**a**) Antenna radiation area. (**b**) Setting of antenna and tag

of the moving person from the antenna to tag. As a result, the person's moving area around the antenna is the key factor of $|\mathbf{V}_M|$. The RFID antenna is usually directional and has an effective radiation area, inside which a 3 dB beamwidth area (denoted as 3 dB-area) exits. Figure 5.19a shows a 3 dB-area for the Laird antenna. The area inside the red circle is the effective radiation range, and the inner area segmented by the two black arrows is the 3 dB-area. When the person moves inside the 3 dB-area, more multipath signals are reflected by the person with a stronger signal magnitude and vice versa.

To see the effect of people's moving area around the antenna on the signal phase, we attach an RFID tag on a stationary box and place the antenna 1.5 m away facing to the tag straightly, as shown in Fig. 5.19b. A red line is drawn on the ground as the 3 dB beamwidth boundary. Volunteers are asked to walk insides, outside, and randomly in and out of the 3 dB-area without blocking the LOS path, as shown in Fig. 5.20a. Since volunteers move closer to the antenna, different moving distances to the tag, i.e., the effect from the uplink signal, introduce limited impact to the signal phase and can be ignored. To compare the phase changes caused by the respiration activity and those brought by the surrounding movements in different areas, we employ the standard deviation (std) of the signal phase, which can serve as a good indicator to measure the variations in the signal. The larger the std is, the larger phase changes are incurred by the movement. The distributions of the signal phase's std for people moving inside, outside, and randomly in and out of the 3 dB-area are shown in Fig. 5.21. We also depict the std of the signal phase merely caused by the respiration activity. From Fig. 5.21, we obtain the following observations: (1) The std of the signal phase when people move inside 3 dB-area is generally larger than that of outside the 3 dB-area. This is because the movements inside the 3 dB-area can result in a larger $|\mathbf{V}_M|$. (2) The std distributions of the random moving and respiration overlap each other, showing that the multipath signals of moving people have similar effects on the phase changes compared with the respiration activity. Therefore, surrounding people's movements could bring comparable phase changes as respiration activity.

For propagation way (2), the moving person mainly affects the uplink of multipath signals, i.e., [tag $\rightarrow$ person $\rightarrow$ antenna]. In this case, $|\mathbf{V}_M|$ is mainly decided by the strength of multipath signals, which are dominated by the reflection of the moving person from the tag to antenna. Therefore, the distance between the moving person and tag becomes the key factor for the phase changes. To observe this

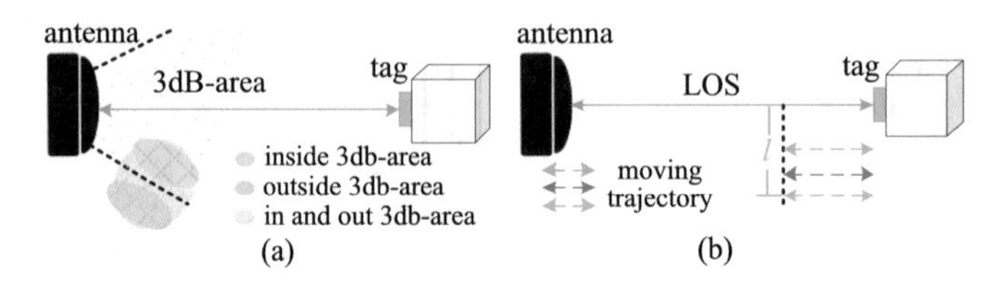

Fig. 5.20 Moving area and trajectory of the surrounding person. (**a**) Move around antenna. (**b**) Move around tag

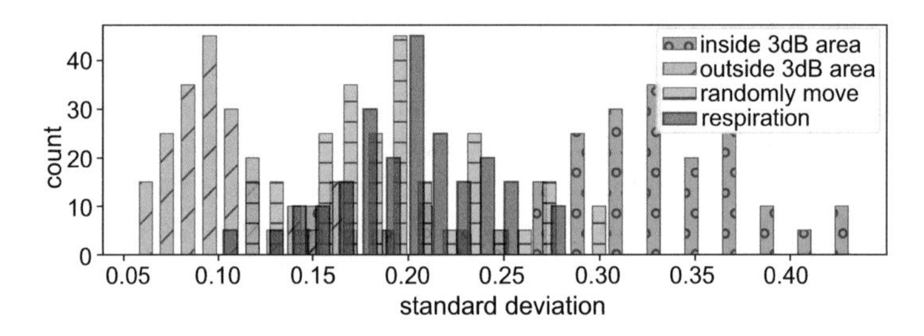

Fig. 5.21 Distribution of standard deviation of the signal phase with moving people moving in different areas

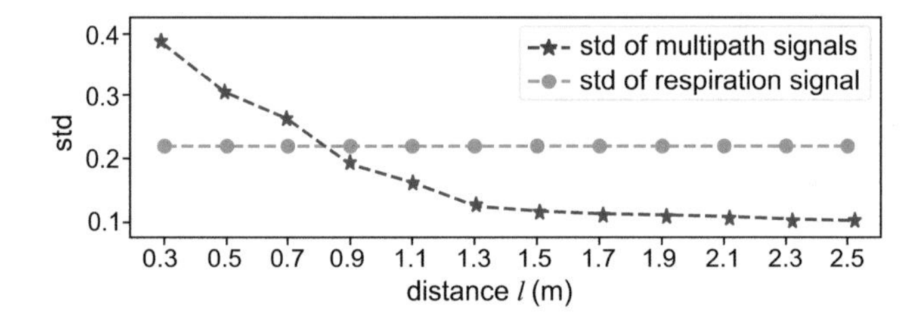

Fig. 5.22 Standard deviation of signal phase for different distances l

effect, we ask a volunteer to walk along a straight line nearby the tag with different distances l to the LOS line, as shown in Fig. 5.20b. Since the volunteer mainly moves around the tag and is relatively far from the antenna, different moving areas towards the antenna, i.e., the effect from the downlink signal, cause little effect on the signal phase and can be neglected. The average std of the signal phase for different l is given in Fig. 5.22. The std first falls sharply and then decreases smoothly along with the increase of l. Thus, the effect of multipath signals when the moving person is far from the monitored person is limited. However, if the person moves close to the monitored person, multipath signals could affect the respiration pattern and should be carefully removed.

Effect of $\angle \mathbf{V}_M$

The effect of $\angle \mathbf{V}_M$ can be revealed from the moving pattern of surrounding people. We analyze the people's moving pattern from two aspects. First, the torso movement can result in two possible changes of $\angle \mathbf{V}_M$, i.e., rotating clockwise and counter-

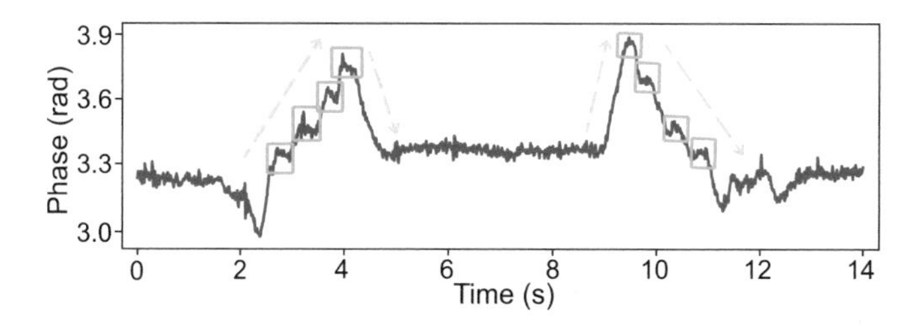

Fig. 5.23 Phase of multipath signals caused by a person walking by

clockwise, which causes the increase and decrease of $\angle \mathbf{V}_M$. Second, people's limbs could swing periodically during walking, which could lead to a rhythmic change of $\angle \mathbf{V}_M$ whose frequency is similar to the limb swing frequency within the range of 1.5–2.5 Hz [14]. Thus, people's moving also brings relatively high-frequency components in the signal phase compared with the human respiration frequency range of 0.17–0.55 Hz.

To show the effect of surrounding people's torso and limb movements, we fix the tag on a stationary box and ask a person to walk from the antenna towards the tag, then stop for a while, and finally walk backward. The measured signal phase is depicted in Fig. 5.23. The general increasing and decreasing trend (highlighted by yellow dashed arrows) are mainly caused by the torso moving from the antenna to the tag side. Meanwhile, the small peaks in the green rectangular are due to the periodic limb movements during walking. The effect from the high-frequency limb movements can be removed by a low pass filter. However, the general increasing and decreasing trend in Fig. 5.23 may be mis-detected as fake respiration cycles, which should be eliminated from the signal phase.

In sum, based on the analysis of $|\mathbf{V}_M|$ and $\angle \mathbf{V}_M$, the multipath signals of moving people could distort the respiration signal with comparable magnitude changes of the signal phase, which include both high-frequency noises and fake respiration cycles.

5.1.2.2 Respiration Signal Mixed with Multipath Signals

To investigate the impact of multipath signals on the respiration signal, we attach an RFID tag on a person's chest and ask another two persons to walk nearby. The received signal phase, which is mixed with respiration and multipath signals, is shown in Fig. 5.24a. The ground truth signal of respiration is collected with a chest band and shown in Fig. 5.24b. The monitored person is asked to breathe normally for 5 respiration cycles. In Fig. 5.24a, the respiration cycles are messed up with noises caused by multipath signals. In particular, the noises in the green rectangular exhibit a similar magnitude as the real respiration peaks. If a low pass filter is applied on Fig. 5.24a followed by a peak detection scheme, as shown in Fig. 5.24c, 7 respiration cycles will be detected, and the extra 2 fake respiration cycles could lead

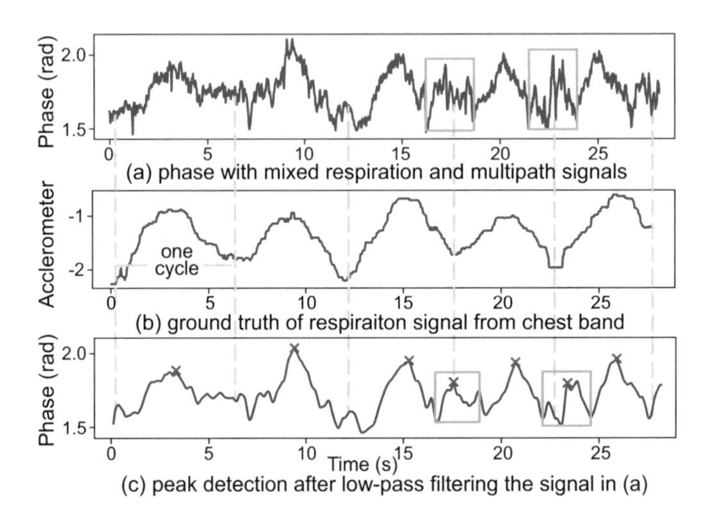

Fig. 5.24 (**a**) Signal phase of multipath mixed respiration signal, (**b**) ground truth respiration signal, and (to the limb swing frequency within) peak detection result on the smoothed signal phase

to inaccurate respiration rate estimation. Besides, multipath signals would cause wrong apnea detection. If people are moving around the monitored person with the apnea syndrome, multipath signals will incur a similar pattern as respiration, which could misguide that the monitored person is still breathing.

5.1.2.3 Apnea Detection

As we discussed in the previous subsection, multipath signals from moving people could result in fake respiration cycles, which can lead to mis-detection of apnea. For example, the signal phase shown in Fig. 5.25a is collected from a person who stops breathing from 15–24 s with people moving around. The multipath signals result in a respiration-like peak from 15–20 s in the signal phase after the low-pass filter, as shown in the red rectangular of Fig. 5.25b. Then, the person would be mis-detected as breathing normally after applying the peak detection scheme on the filtered phase.

To differentiate the apnea from multipath signals, we employ the time-frequency pattern of the signal phase. In specific, we extract the spectrogram of the signal phase, from which we can identify the anomaly in the respiration signal with apnea. For instance, the spectrogram of the signal phase in Fig. 5.25a is extracted and shown in Fig. 5.25c. The signal spectrogram exhibits a white area in the middle which exactly matches the apnea period, meanwhile clearly showing the dominant frequencies during normal breathing at around 0.3–0.4 Hz, which corresponds to the respiration frequency. This indicates that the frequency components of the real respiration signal, although mixed with the multipath signals, are still dominant over the respiration frequency range of 0.17–0.5 Hz. In contrast, if the person stops

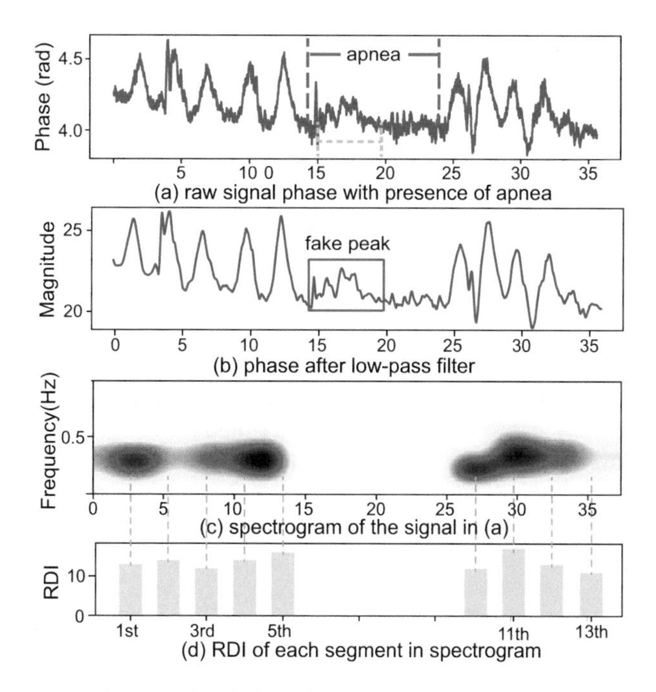

Fig. 5.25 Apnea signal: (**a**) raw signal phase, (**b**) signal phase after matched filter, (**c**) spectrogram, and (**d**) RDIs of the spectrogram

breathing and only multipath signals are left, the frequency components almost disappear within the respiration frequency range.

Based on this observation, we leverage the disappearance of the dominant frequencies within the respiration frequency range to detect the apnea. We define a respiration-dominance index (RDI) to detect whether the dominant frequency disappears within the respiration frequency range. To measure RDI, we first perform the short-time Fourier transformation (STFT) on the signal phase to obtain the spectrogram. In STFT, the signal phase is first divided into fixed-length segments.[1] For each time segment, we measure the mean of all the frequency-domain amplitudes in the spectrogram as a noise threshold. Then, the RDI is calculated by counting the number of frequencies, whose amplitude exceeds the noise threshold within the respiration frequency range. RDI characterizes the dominance of the respiration components over all the frequency components in each segment of the signal phase. The RDI of multipath signals during the apnea period is much lower than that of the respiration cycles mixed with multipath signals. In Fig. 5.25d, we show the RDIs for all the segments of the spectrogram in Fig. 5.25c. RDIs during the apnea period all decrease to 0, while the RDIs for the respiration cycles all exceed 10. To detect the

[1] In our implementation of STFT, the length of the segment is set to 512 sampling points, and the size of FFT is 2048 after zero-padding.

decrease in RDIs for apnea detection, we calculate the mean RDI of the person's pre-collected respiration signal in the static environment, and half of the mean RDI is set as the reference RDI. If the length of consecutive RDIs whose values are lower than the reference RDI exceeds 5 s, the apnea is detected, and the corresponding phase window will not be used to perform respiration rate estimation. The length of 5 s is chosen because it is the longest duration of normal respiration cycles.

5.1.2.4 Matched Filter

After apnea detection, the signal phase without apnea is used to estimate the respiration rate. Recall that the real respiration cycles are distorted by high-frequency noises and fake respiration cycles, which cannot be simply eliminated by the low-pass filter. To tackle this issue, we leverage the difference between the respiration pattern and multipath signals. In specific, due to the intrinsic features of the human respiration pattern, the respiration signal phase shows a periodic and sinusoidal pattern. In contrast, multipath signals are random and irregular, which involve both low and high-frequency noises combined in various ways. This inspires us to employ the matched filter to detect the target signal out of noises. The matched filter is an optimal linear filter, created from a target signal template, to detect the target signal by maximizing its signal-to-noise ratio (SNR) from the unknown signal mixed with noises [26]. For respiration monitoring, we extract a single respiration cycle as the template for creating the matched filter and apply the matched filter on the received signal phase to denoise it. The output of the matched filter will peak at where the target signal appears. Finally, we can detect the peaks in the filtered phase and estimate the respiration rate.

To design the matched filter, the respiration cycle template should be carefully selected due to the following reasons. First, the shape of the template can affect the performance of the matched filter. Only when the template has the same shape as the target signal can we achieve the optimal SNR. If the shape of the template is not consistent, the SNR of the matched filter output will vary accordingly. Second, respiration patterns are unique and diverse among different people [5]. For example, the signal phase of two persons' respiration in Fig. 5.26 shows that their respiration cycles have different shapes. This means that the respiration cycle template should be typical for each monitored person to achieve higher SNR of the filtered phase.

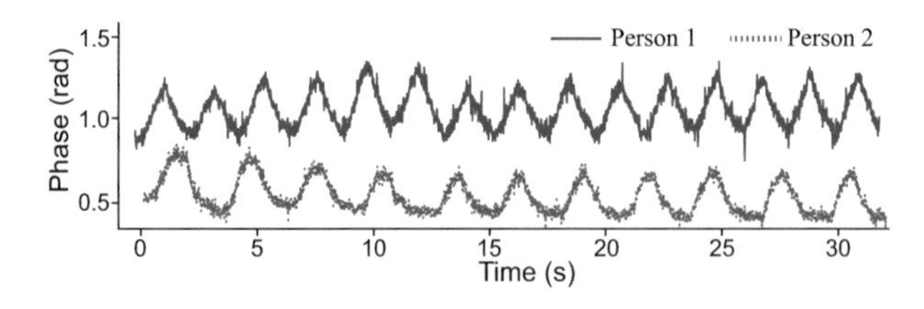

Fig. 5.26 Respiration signals of two persons, showing different patterns

To extract the respiration cycle template, we first pre-collect the signal phase of pure respiration for the monitored person in a static environment. The monitored person normally breathes for 1–2 min during which the signal phase is collected. Noth that the template collection is a one-time step, which would not bring too much inconvenience to users. Then, we extract the template from the pure respiration signal by using a cycle-averaging method introduced as follows. First, we smooth the respiration signal phase with a median filter. Then, we detect the local minimums, which are the starting points of respiration cycles, to segment the signal phase into individual cycles. To detect the local minimums, peak detection is performed on the negative of the signal phase. Next, for each respiration cycle, we calculate its similarity with all the other respiration cycles using the Euclidean distance. The respiration cycle with the highest similarity is selected as the template candidate. Finally, the template candidate is scaled according to the average width and height of all the respiration cycles as the respiration cycle template $r_t(n)$.

With the extracted template $r_t(n)$, the impulse response of the matched filter $h(k)$ is obtained as $h(k) = r_t(N - k - 1)$, where N is the length of $r_t(n)$. In Fig. 5.27, we show the output signal phase after applying the matched filter on the raw signal phase in the upper figures of Fig. 5.27a, b, respectively. In the first figure of Fig. 5.27a, the multipath signals from surrounding movements bring fake respiration cycles in the raw signal phase. When using the low-pass filter, these fake cycles still remain in the signal phase. In contrast, applying the matched filter can remove the fake cycles meanwhile accurately detecting the real cycles, which match the ground truth in Fig. 5.24b. Similarly, by applying the matched filter, the fake respiration peak caused by the limb movement in Fig. 5.27b, is successfully removed, which, however, cannot be fulfilled by the low-pass filter.

5.1.2.5 Respiration Rate Estimation

Intuitively, we can apply FFT to measure the respiration rate. However, the resolution of FFT is restricted by the length of the time window. For instance, if respiration rate is measured every 20 s, the resolution in the frequency domain is 0.05 Hz, which results in 3 bpm resolution in the time domain. Thus, to accurately estimate the respiration rate, we use peak detection to avoid the low-resolution problem of FFT.

The peak detection method estimates the respiration rate based on the detected peaks, which is suitable for real-time respiration monitoring. However, the peak detection approach could suffer from tiny fluctuations, which can be misdetected as peaks in the filtered phase. Previous methods set thresholds to discard the wrong peaks which are too low or too closed to each other [17]. However, in the RM scenario, the magnitude of the signal phase will change along with time. A fixed threshold may be improper and could incur missing or wrong peaks. Therefore, to adapt to different scales automatically, we employ the automatic multi-scale peak detection (AMPD) [28] algorithm. AMPD frees us from choosing fixed thresholds to detect the real peaks with the help of the multi-scale technique. The detected

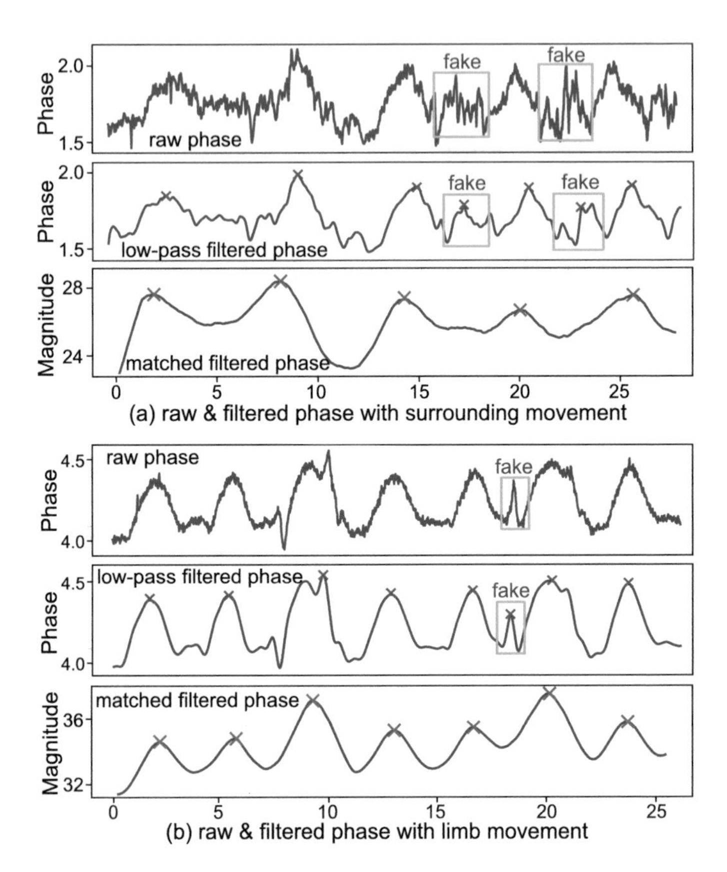

Fig. 5.27 Raw and filtered phase mixed with the multipath signals from (**a**) the ambient movement of surrounding people and (**b**) the limb movement of the monitored person. The matched filter can effectively remove the fake respiration cycles and noises

peaks of the signals in Fig. 5.27 after applying AMPD are shown with red crosses. Then, the respiration rate is estimated as follows:

$$rate = 60/\frac{1}{n}\sum_{i}^{n-1}(p_{i+1} - p_i), \tag{5.14}$$

where p_i is the timestamp of the detected peak, and n is the total number of peaks. The calculated respiration rate is in the unit of breath per minute (bpm).

5.1.2.6 Evaluation

We introduce the experimental setup, evaluation metrics, and experimental results in terms of different factors for apnea detection and respiration rate estimation.

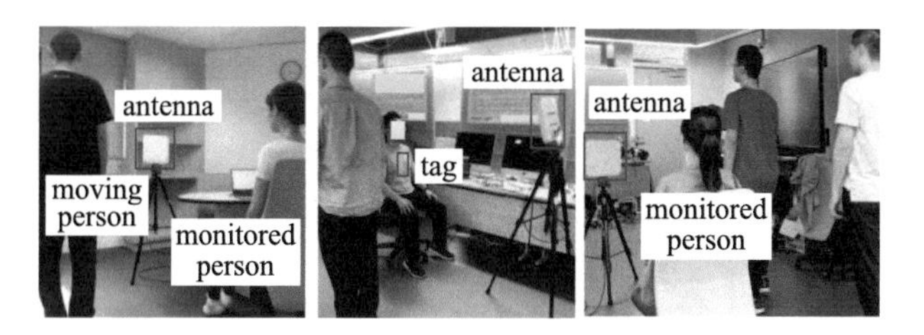

Fig. 5.28 Experimental settings in different environments

Experimental Setup We implemented the system using commercial off-the-shelf RFID devices. The ImpinJ Speedway R420 reader is connected with a Laird E9208 antenna to transmit the RFID signal and interrogate the RFID tag. The reader works in the 920–925 MHz frequency band, and the reader mode is set to MaxThroughput. The reader is connected to a Dell Inspiron 7460 laptop with i7-7500U CPU and 8 GB RAM. The RFID signal phase is processed using Python 3.0. We conducted experiments in three different environments with different layouts, as shown in Fig. 5.28. The antenna is placed 1–2 m away from the monitored person. The tag is attached on the person's chest. We invite 12 volunteers, including 3 females (height: 165–170 cm, chest width: 27–32 cm) and 9 males (height: 172–180 cm, chest width: 33–40 cm), to act as the monitored person and surrounding people in turn. We do not assign specific routes for volunteers to move so that they can walk freely nearby the monitored person. We ask the monitored person to normally breathe for 2 min to extract the respiration template. Then, the signal phase is segmented into 20 s-windows to estimate the respiration state. The ground truth of the respiration signals is collected via a chest band equipped with a 3-axis accelerometer.

Evaluation Metrics We use the following metrics to evaluate the performance of our system. First, for apnea detection, the percentage of the missing apnea (MA) and false apnea (FA) over all the apnea cases are defined as follows:

$$MA = \frac{\#missing\ apnea}{\#real\ apnea},\quad FA = \frac{\#false\ apnea}{\#no\ apnea}. \tag{5.15}$$

Second, to evaluate the accuracy of respiration rate estimation, we use the mean absolute error (MAE) as follows:

$$MAE = \frac{1}{n}\sum_{i=1}^{n}|r_i - r_i^{'}|, \tag{5.16}$$

where r_i and $r_i^{'}$ are the estimated and real respiration rate, respectively. n is the number of time windows.

Table 5.1 Comparison of RDI-based and peak-threshold based methods for apnea detection

	RDI (our method)	Peak-threshold
MA	3.75%	12.65%
FA	4.15%	15.8%

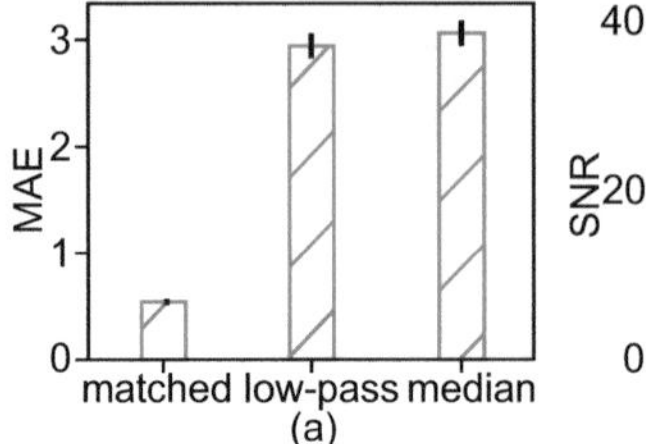

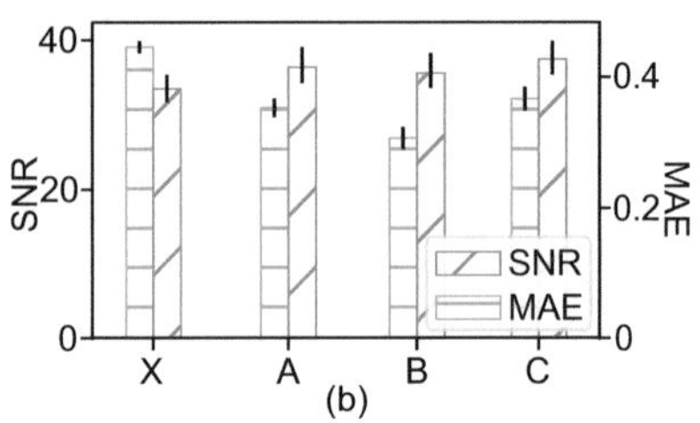

Fig. 5.29 Effect of matched filter: (**a**) MAE for respiration rate estimation with matched filter, low-pass filter, and median filter, (**b**) SNR and MAE of using different respiration templates to make matched filter

Effect of RDI for Apnea Detection In this evaluation, we compare the performance of our RDI-based approach with the existing peak-threshold approach [34] to demonstrate the effectiveness of our approach on apnea detection in dynamic environments. The previous approach sets a fixed threshold for detecting peaks in the respiration signal. If there is no peak for a certain time, the apnea is detected. We set the same threshold in [34], which is the median of the signal phase in a time window, and compare its result with our RDI-based method. Table 5.1 shows that our approach outperforms the peak-threshold approach with an approximate 10% reduction of MA and FA. This is because the fake peaks caused by the multipath signals from moving people are wrongly regarded as breathing cycles in the peak-threshold approach. However, our proposed RDI-based approach can accurately differentiate the real respiration signal from the multipath signals during the apnea period.

Effect of Matched Filter for Respiration Rate Estimation To show the effectiveness of the matched filter on respiration rate estimation, we first compare the MAE between the real and estimated respiration rates with and without applying the matched filter on the signal phase. For methods without the matched filter, we use the low-pass filter and median filter to denoise the signal phase. Then, AMPD is applied to count the respiration cycles. As shown in Fig. 5.29a, the average MAE using the matched filter is 0.51. While the MAEs using the low-pass and median filters are 2.94 and 3.08 bpm, respectively, which are 5 times larger than that of using the matched filter. This indicates that the matched filter can help to promote the accuracy of respiration rate estimation.

Next, we investigate the effectiveness of the template extraction and update methods with two experiments. The first experiment is to show how different persons' templates could affect the RM performance. We select one volunteer (X) and extract the template from X's respiration signal phase in a static environment.

Then, X's template is used to create a matched filter to denoise the respiration signal phase when people move around X. Then, we extract another three templates from three volunteers (A, B, and C) and create three matched filters, respectively. Finally, we apply the three matched filters to denoise X's respiration signal phase. The MAEs of using the matched filters created from different persons' templates are shown in Fig. 5.29b. In addition to the MAE, SNR is also reported to show the ability of the matched filter for denoising the signal. The SNR when using the matched filter created from X's own template is higher than using other persons' templates. Meanwhile, using X's own template also achieves the lowest MAE, indicating the importance of extracting the personalized template for each user and the effectiveness of our template extraction method.

The second experiment is to show the performance of the template update method when the monitored person changes the respiration pattern. We let X breathe normally for 5 min with other people moving nearby and collect the respiration signal phase. Then, X is asked to do pedaling for 15 min. After pedaling, the respiration rate of X greatly increases, and we continue to collect X's signal phase for 10 min. Then, we estimate the respiration rate before and after pedaling using a fixed template and the updated templates of X, respectively. The template update period is set as 2 min. The MAE of using the update templates is 0.11 bpm lower than the fixed template, showing the effectiveness of the template update method.

Effect of the Number of Moving People on Apnea Detection and Respiration Rate Estimation In this evaluation, we evaluate the system performance in both static and dynamic environments with different numbers of moving people, i.e., 0 (static), 1 (1p), 2 (2p), and 3 (3p) persons moving around. The distance between the person and antenna is set as 1.5 m. First, we evaluate the apnea detection performance. Volunteers who act as the monitored person are asked to simulate the apnea by holding their breath for 5–10 s. The results of MA and FA are shown in Fig. 5.30a. Generally, MA and FA grow slightly with the increasing number of moving people, while they are both below 6% for all cases. The average MA and FA with 1–3 moving persons are only 2–3% higher than those in the static environment. This indicates that our approach can enhance the robustness of apnea detection in dynamic environments with multiple surrounding persons.

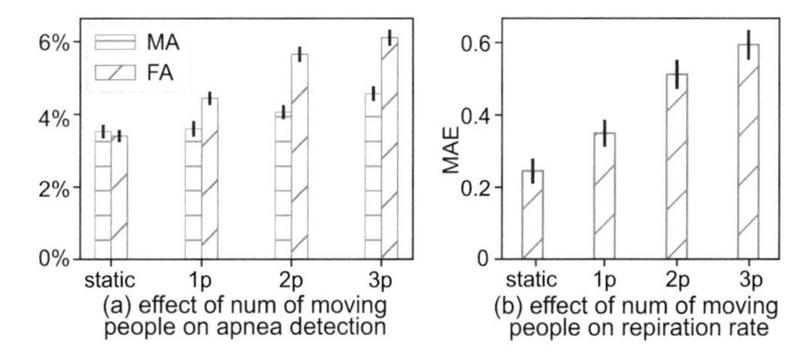

Fig. 5.30 Effect of the number of moving people on (**a**) apnea detection and (**b**) respiration rate estimation

5.1.3 RFID-Based Concurrent Exercise and Respiration Monitoring

Monitoring physical and physiological states during exercise is of vital importance for people's health and safety during exercise. Many people fail to achieve safe and effective exercise due to the lack of measurements for the physiological signals, which could result in serious physical injuries or visceral organ damage. Various parameters have been used to monitor the exerciser's physical and physiological status during exercise. One key parameter is the locomotor-respiratory coupling (LRC) ratio, which characterizes the correlation between the exercise locomotion and respiration rhythm. Researchers have found that there exists a tight coordination between the limb movement and respiration for many rhythmic exercise activities, e.g., running, cycling, and pedaling [6]. Keeping harmonic coordination between locomotion and respiration during exercise not only reduces energy consumption and prolongs the training period, but also promotes the maintenance of the cardiopulmonary functions [19]. LRC ratio is an essential parameter, which is defined as the frequency and phase locking between locomotion and respiration [4]. By monitoring the LRC during exercise, people can adjust their respiratory rhythm to maintain the LRC ratio at a stable and proper level, so that they can boost exercise performance and prevent getting hurt.

Estimating LRC requires simultaneous monitoring of the exercise and respiration activity. However, most existing wireless-based systems can only monitor exercise or respiration separately. For example, WiFi [40] and RFID [10]) are used to detect and evaluate the exercise performance for various exercise activities. WiFi [16] and RFID [32] are also used to monitor the respiration state but only when the person is in the quasi-static state, e.g., sleeping and sitting. However, it is a challenging task to achieve respiration monitoring during exercise using the wireless signal. This is because the chest movement during respiration is too subtle to detect in face of the overwhelming effect from the large-scale exercise movement.

In this case study, we showcase how commercial RFID devices can be employed to measure the LRC ratio. RFID tags are attached on the limbs and chest to extract the exercise limb movement and respiration rhythm, respectively. As shown in Fig. 5.31, two RFID antennas are deployed in front of and behind the person. The

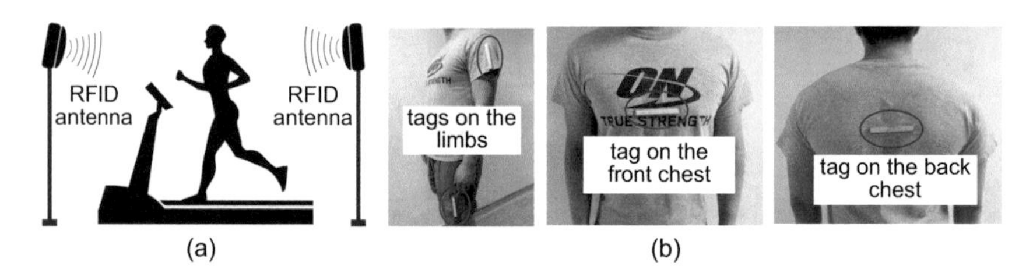

Fig. 5.31 Illustration of the system scenario and deployment of RFID tags. (**a**) System deployment. (**b**) Deployment of the RFID tags on the body

Table 5.2 Common LRC ratios for human-beings during exercise

LRC ratio	5:1	4:1	3:1	5:2	2:1	5:3	3:2	4:3	1:1

front antenna transmits the signal to the tags on the limbs and front chest. The back antenna interrogates the tag on the back chest. The backscattered signals reflected by all the tags are collected by the RFID reader for further processing. To accurately estimate the LRC, we first study the feasibility to measure both the exercise limb movement and respiration rhythm by modeling the effects of different body movements on the phase values of the RFID signal. Then, we estimate the exercise limb movement rate with the limb tags. The tiny respiration pattern is extracted after removing the effects from the large torso movement by synthesizing the information of multiple tags on the body. Finally, the patterns measured from the limb movement and respiration are coupled to estimate the LRC ratio.

5.1.3.1 Understanding the Exercise and Respiration Rhythm

LRC is defined as the frequency and phase locking between the locomotor and respiratory systems [29]. According to the LRC theory [7], for humans doing exercise, there are a set of coupling ratios as listed in Table 5.2, denoted as LRC list. Taking the LRC of 3:1 for the cycling activity as an example, it means the person breathes 1 time with 3 rounds of periodic limb movement. Among the LRC list, integer LRC ratios appear more frequently than non-integer ratios [6]. The type of exercise activity is one of the modulators of the LRC. For example, LRC ratios in running are typically 2:1, 3:1, and 4:1, among which 2:1 is the most common ratio. For cycling, common LRC ratios are 2:1 and 3:1, while 1:1 is more frequently present in weightlifting. The personal trainers usually advise people to breathe within the common range of LRC for the specific activity. Exercisers with higher LRC may overload themselves without sufficient oxygen supply. While lower LRC indicates that exercisers may achieve less effective exercise training.

The way to measure LRC, which has been adopted by many researchers, is to divide the locomotion movement rate by the respiration rate during a period and approximate the result to its nearest ratio in the LRC list [23]. There can be few cases, e.g., when the real LRC is 11:4, while the determined LRC is 3:1. However, according to the studies in sports training [19] and interview replies from personal trainers, it is sufficient to provide the common LRC ratios in the LRC list to exercisers for training guidance. Hence, the measured LRC is one of the values in the LRC list.

5.1.3.2 The RFID Signal of Exercise or Respiration Movements

The phase value of the RFID signal is used to monitor the exercise and respiration activity. The phase value is a basic metric that reveals the relative distance information between the signal transmitter and receiver. For the RFID signal, the distance d between the antenna and the tag can influence the phase value φ, which can be represented as follows:

$$\varphi = \left\{ \frac{2d}{\lambda} \cdot 2\pi + \epsilon \right\} \mod 2\pi, \tag{5.17}$$

where λ is the RFID signal wavelength ($\approx$ 32.4 cm here), and ϵ is the phase shift with a fixed value caused by the hardware imperfection and environmental effects. Since we focus on the change of the phase values during body movements, the existence of ϵ will not affect the result. When the tag is attached on the body, it will move along with the body during exercise, and the phase values will change accordingly. The limb movement can bring around 5–15 cm change on d depending on the position of the tag on the limb. While the chest movement during respiration only incurs about 5–10 change on d, so the phase changes incurred by the respiration would be quite small.

The phase values of the tags on the limbs are shown in Fig. 5.32a, b for the pedaling and running activities as examples. They show clear and periodic patterns of limb movements. Then, a person is asked to breathe normally in a quasi-static standing state, and the phase values of the tag on the front chest are shown in Fig. 5.32c, in which 5 times of sinusoidal breathing cycles are present. The phase value for respiration is much smaller than the limb movement. Furthermore, the respiration signal of the chest tag during exercise would suffer from more noises due to the effects of the large torso movement.

5.1.3.3 Modeling the Respiration Activity During Exercise

In this subsection, we analyze the effects of limb movement and respiration on the RFID signal. The tags on the limbs, i.e., the leg and arm, are denoted as T_l, and the tags on the chest are represented as T_c. For T_l, it is mainly affected by the limb movement effect (E_l), which is the cyclic movement of arm and leg, as shown in Fig. 5.33a. The phase value φ_l of T_l can be expressed as follows:

$$\varphi_l = \left\{ \frac{2(d_l + d_{E_l})}{\lambda} \cdot 2\pi \right\} \mod 2\pi, \; d_{E_l} = A_1 \sin(2\pi f_l + \varphi_1). \tag{5.18}$$

d_l is the initial distance between the tag and the antenna. d_{E_l} characterizes the periodic limb movement into a sinusoidal signal, with f_l as the limb movement rate and φ_1 as the initial phase.

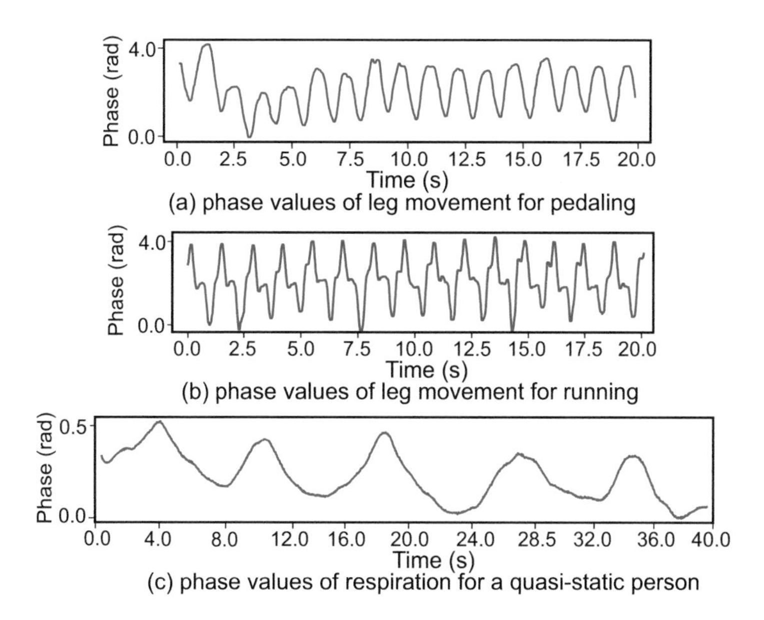

Fig. 5.32 Phase series of limb movements for exercise activities and chest movement for respiration: (**a**) signal phase during pedaling, (**b**) signal phase during running, (**c**) signal phase during respiration

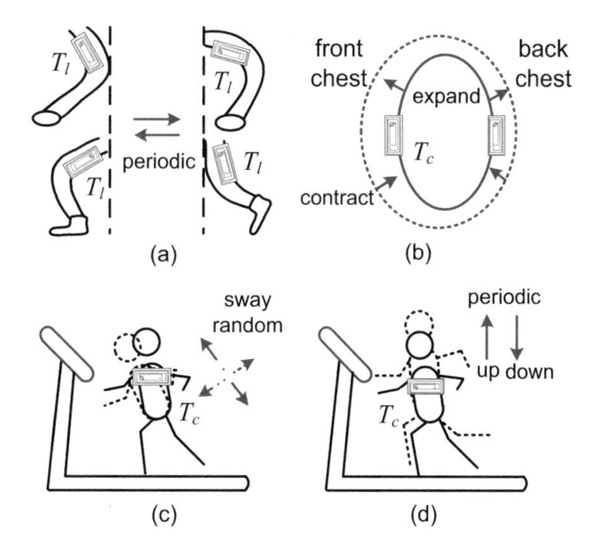

Fig. 5.33 Different movement effects on the tags for the running activity. (**a**) limb movement effect E_l. (**b**) Respiration effect E_r. (**c**) random effect $E_t r$. (**d**) period effect $E_t p$

For T_c, it is directly influenced by two sources of movement effects via the LOS path, as shown in Fig. 5.33b–d, one is the target respiration effect (E_r) caused by the chest movement for observing the respiration pattern during exercise (Fig. 5.33b), the other is the torso movement effect (E_t). The torso movement effect can be further divided into two parts: random effect (E_{tr}) and periodic effect (E_{tp}). The random effect is caused by the random movement during exercise. Although most indoor exercise activities are in-place movements, the torso could move randomly to the right or the left, forward or backward (Fig. 5.33c), as the body cannot keep definitely balanced while exercising. The periodic effect is the product of the cyclic torso movement during exercise. For example, the person's torso would sway up and down while running on a treadmill (Fig. 5.33d). As a result, the periodic effect could synchronize with the limb movement effect. The torso movement effect could degrade the respiration pattern in the phase values. This is because that the torso movement is larger than the minute respiration movement. Thus, the respiration pattern can be overwhelmed by the presence of torso movement. There are also indirect and minor multipath effects brought by the limb movement on T_c, denoted as $d_{E_{lm}}$, which is consistent with the limb movement effect. Thus, the phase values φ_c of T_c can be represented as follows:

$$\varphi_c = \left\{ \frac{2(d_c + d_{E_r} + d_{E_t} + d_{E_{lm}})}{\lambda} \cdot 2\pi \right\} \mod 2\pi,\ d_{E_r} = A_2 \sin(2\pi f_r + \varphi_2).$$
$$d_{E_t} = d_{E_{tr}} + d_{E_{tp}},\ d_{E_{tp}} = A_3 \sin(2\pi f_l + \varphi_3),\ d_{E_{lm}} = A_4 \sin(2\pi f_l + \varphi_4). \tag{5.19}$$

The respiration effect d_{E_r} is also converted into a sinusoidal signal, with f_r as the respiration rate. Since the indirect multipath signals from the limb movement are weaker than the direct period effect, thereby $A_4 < A_3$.

To extract the exercise locomotion pattern, we can leverage the phase values of T_l and estimate the limb movement rate f_l via peak detection. However, the large torso movement effect, mixed with the respiration effect, should be removed from the signal of T_c to recover the clear respiration pattern. Figure 5.34a depicts the raw RFID phase values of the tag on the front chest during pedaling, which are quite messy to see the periodic respiration pattern due to the existence of the torso movement effect. The large wave changes in the signal are brought by the random effect, and the small and high-frequency fluctuations are caused by the periodic effect. Figure 5.34b shows the recovered respiration signal with clear and periodic respiration cycles after our dedicated processing, which will be introduced in Sect. 5.1.3.5.

5.1.3.4 Limb Movement Rate Estimation

The limb movement rate f_l is estimated from the signal of the tag on the limb. For running, cycling, and pedaling, f_l is estimated with the tag on the leg. For weightlifting and rowing, f_l is obtained with the tag on the arm. The limb movement

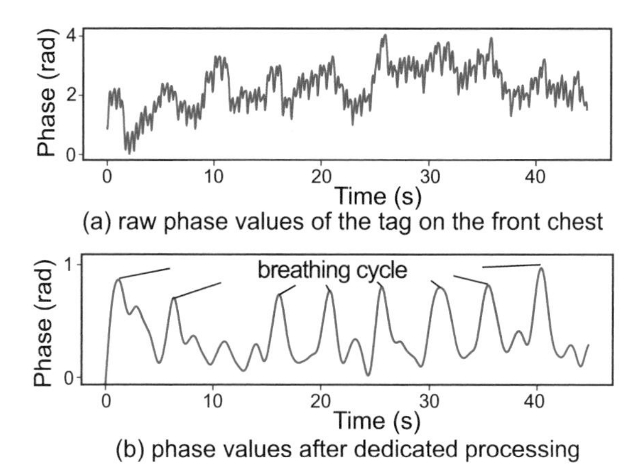

Fig. 5.34 (**a**) raw and noisy signal phase of the tag on the front chest, (**b**) the signal phase after processing, which shows a clear respiration pattern

pattern in Fig. 5.32a, b presents a sinusoidal wave which inspires us to detect the limb movement cycle via peak detection. However, since the peak detection algorithm identifies the peak whose neighbor value is smaller than it, there are many fake peaks in the phase series as shown in Fig. 5.35a. To remove the fake peaks that do not correspond to the real movement cycle, we set a threshold for the distance between two neighbor peaks. For running, the stride frequency is usually in the range of [90, 120] cycles per minute. For cycling and pedaling, the number of revolutions is within 80 rounds per minute. For rowing and weightlifting, the number of cycles could be less, which is below 40 times per minute. Therefore, we can calculate the minimum interval σ_t between two movement cycles and discard the peaks that are too close to each other. In addition, we only keep the peaks whose height are higher than the mean value σ_m of the phase series. In Fig. 5.35b, the peaks that represent the locomotion cycles are detected. Then, the number of peaks is counted in a time window t as p_n. Since f_l is relatively stable within a short period of time, the limb movement rate f_l is estimated by $f_l = p_n/t$.

5.1.3.5 Respiration Pattern Extraction

Next, we will extract the respiration pattern involved in the phase values of T_c via respiration effect measurement and torso movement effect removal.

Respiration Effect Measurement

We employ the respiration movement mechanism, as shown in Fig. 5.36a, to fully investigate and measure the effects of chest displacement on the RFID phase values. When a person breathes in and out, the whole chest, which is simplified into a cylinder, will expand and contract accordingly. In this process, there would be displacement change in the front, mediolateral, and back dimensions of the chest. During exercise, the mediolateral dimension change could be buried by the arm movement. While the front and back dimensions can be fused together to extract

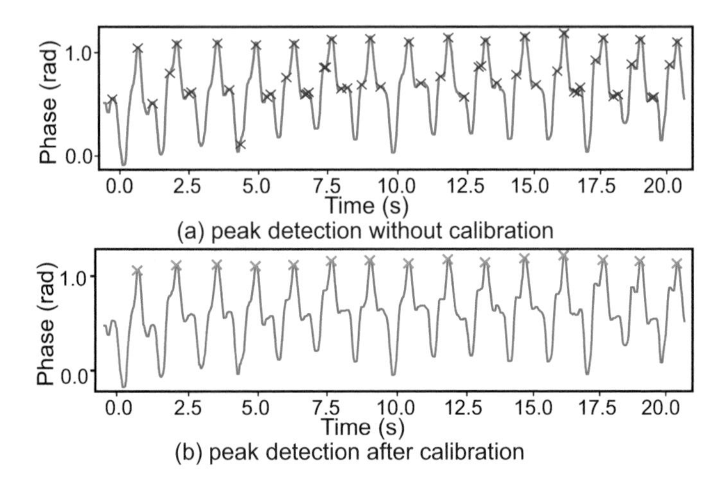

Fig. 5.35 Limb movement rate extraction with peak detection: (**a**) with fake peaks, (**b**) after peak calibration

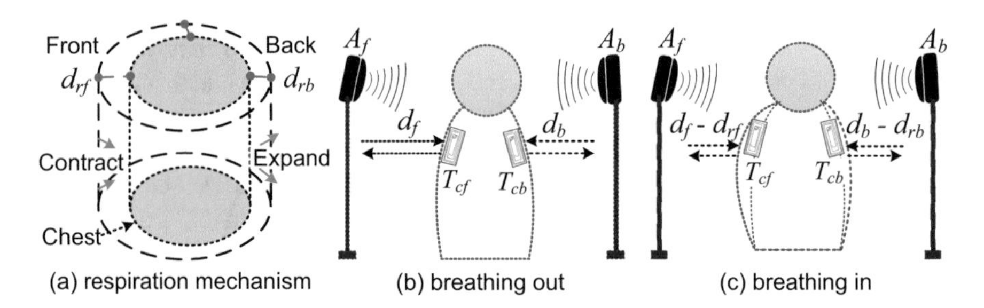

Fig. 5.36 Chest movement modeling during respiration and the propagation of the RFID signal: (**a**) human respiration mechanism, (**b**) signal propagation when breathing out, (**c**) signal propagation during breathing in

the respiration effect, instead of only exploiting the front dimension change. In this way, the respiration pattern can be enlarged. Therefore, we attach one tag on the front chest and one tag on the back chest. Then, two antennas are put right in front of and behind the target person to interrogate the front tag and back tag, respectively, as shown in Fig. 5.36b–c. Suppose that the respiration can lead to d_{rf} and d_{rb} displacement change of the front chest and back chest, respectively. d_{rf} is commonly larger than d_{rb}.

For the tag on the front chest (T_{cf}), the initial distance between T_{cf} and the front antenna (A_f) is d_f. When breathing in, the front chest will expand and move towards A_f, making the phase values of T_{cf} decrease. When breathing out, the chest will contract which leads to the increase of the phase values of T_{cf}. Then the phase difference $\Delta\varphi_f$ of T_{cf} between breathing in ($(\varphi_f)_{in}$) and breathing out ($(\varphi_f)_{out}$) is

$$\Delta\varphi_f = \left|(\varphi_f)_{in} - (\varphi_f)_{out}\right| = \left|(\frac{2d_f - 2d_{rf}}{\lambda} - \frac{2d_f}{\lambda}) \cdot 2\pi\right| \mod 2\pi$$
$$= \left\{\frac{2d_{rf}}{\lambda} \cdot 2\pi\right\} \mod 2\pi. \tag{5.20}$$

For the tag on the back chest (T_{cb}), the initial distance between T_{cb} and the back antenna (A_b) is d_b. When breathing in, the signal's traveling distance will drop by $2d_{rb}$ with the back chest expanding and moving backward. While for breathing out, the signal traveling distance will increase by $2d_{rb}$ with the back chest contracting and moving forward. Suppose we add the phase values of T_{cf} and T_{cb}, the sum values while breathing in and out can be represented as follows:

$$(\varphi_f + \varphi_b)_{in} = \left\{\frac{2(d_f - d_{rf}) + 2(d_b - d_{rb})}{\lambda} \cdot 2\pi\right\} \mod 2\pi. \tag{5.21}$$

$$(\varphi_f + \varphi_b)_{out} = \left\{\frac{2d_f + 2d_b}{\lambda} \cdot 2\pi\right\} \mod 2\pi. \tag{5.22}$$

Accordingly, the phase difference $\Delta(\varphi_f + \varphi_b)$ between breathing in and breathing out can be expressed as:

$$\Delta(\varphi_f + \varphi_b) = \left|(\varphi_f + \varphi_b)_{in} - (\varphi_f + \varphi_b)_{out}\right| = \left\{\frac{2(d_{rf} + d_{rb})}{\lambda} \cdot 2\pi\right\} \mod 2\pi. \tag{5.23}$$

Comparing Eq. (5.20) with Eq. (5.23), the phase difference of the summation of T_{cf} and T_{cb} is $2d_{rb}/\lambda$ larger than that of the single T_{cf}. In this way, the respiration pattern can be amplified. To verify it, we make one person to breathe in the quasi-static state, and measure the phase values of T_{cf}, T_{cb} and calculate the sum of T_{cf} and T_{cb}, as depicted in Fig. 5.37. It shows that the phase series of T_{cf} and T_{cb} are in-phase with each other. The peaks and valleys in the phase series when breathing out and breathing in are generally synchronized. It also shows that the magnitude of the phase values of T_{cb} is smaller than that of T_{cf}, since $d_{rb} < d_{rf}$. Most importantly, the summation of T_{cf} and T_{cb} shows larger amplitude for the respiration pattern in Fig. 5.37c. Therefore, the phases values of T_{cf} and T_{cb} are added together for measuring the respiration chest movement.

Torso Movement Effect Removal

In Sect. 5.1.3.3, we mention the torso movement effect, including the random effect and period effect, can overwhelm the respiration signal for the tags on the chest. Thus, we need to eliminate the effects caused by the torso movement to capture a clear respiration pattern.

For the random effect, we divide the random torso movements into two parts, one is the vertical displacement (move forward and backward), and the other is horizontal displacement (move to the left and right). First, we show how the vertical

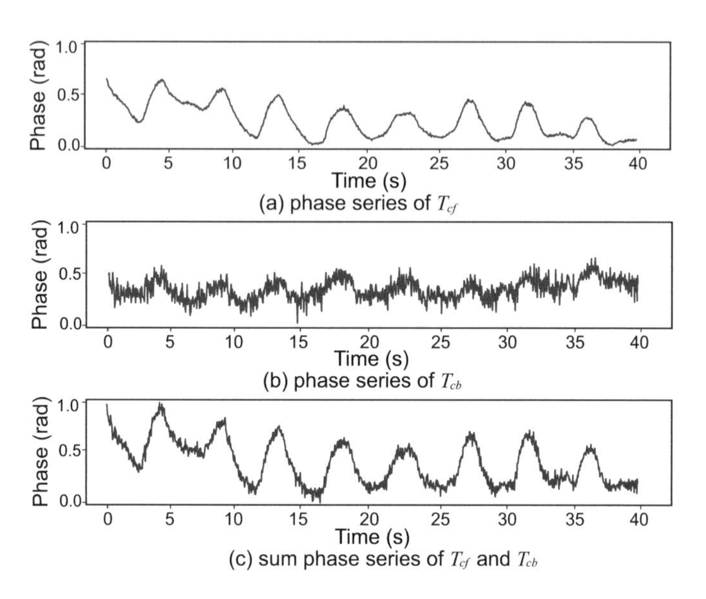

Fig. 5.37 Respiration phase series of T_{cf}, T_{cb} and their sum: (**a**) signal phase of the front tag, (**b**) signal phase of the back tag, (**c**) summation of signal phases of front and back tags

displacement affects the phase values. As illustrated in Fig. 5.38a, the torso moves l forward to the front antenna A_f. Then the phase values of T_{cf} will decrease $\frac{2l}{\lambda}$. Suppose l is 5 cm, as the in-place exercise usually results in a tiny torso swaying movement. Compared with the displacement of the chest while breathing, which is in the range of [5, 10 mm] for the front chest [30], the vertical displacement l could bring a large effect on the phase values. For T_{cb}, the phase values will increase inversely. Next, we discuss the effects of the horizontal displacement. As depicted in Fig. 5.38b, the person moves to the left with distance of $l = 5$ cm. Suppose the distance between T_{cf} and A_f is $d_f = 1$ m, then the phase values will increase $\frac{2(\sqrt{d_f^2+l^2}-d_f)}{\lambda}$. The $(\sqrt{d_f^2+l^2} - d_f)$ is only around 1 mm, which is quite small compared with the effects of vertical displacement $l = 5$ cm, so that the horizontal displacement can be ignored. Therefore, we only need to consider the effects of the vertical displacement. We find that the vertical displacement brings opposite effects on the T_{cf} and T_{cb}. More specifically, if the person does not move, the summation of the phase values of T_{cf} and T_{cb} can be represented as follows:

$$\varphi_f + \varphi_b = \left\{ \frac{2(d_f + d_b)}{\lambda} \cdot 2\pi \right\} \mod 2\pi. \tag{5.24}$$

If the person's torso moves forward l, the propagation path of T_{cf} will drop from $2d_f$ to $2d_f - 2l$, while the path of T_{cb} will increase from $2d_b$ to $2d_b + 2l$. Then, the phase summation of T_{cf} and T_{cb} is as follows:

Fig. 5.38 Modeling of the torso random effect: (a) effect of vertical torso movement, (b) effect of horizontal movement

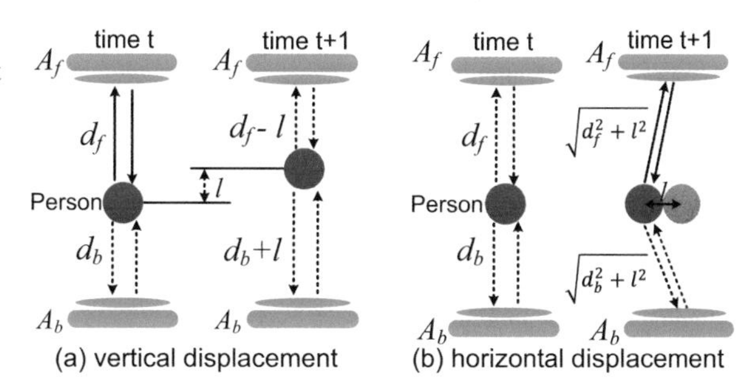

$$\begin{aligned}\varphi_f+\varphi_b&=\left\{\frac{2(d_f-l)}{\lambda}\cdot 2\pi+\frac{2(d_b+l)}{\lambda}\cdot 2\pi\right\}\mod 2\pi\\&=\left\{\frac{2(d_f+d_b)}{\lambda}\cdot 2\pi\right\}\mod 2\pi.\end{aligned}\tag{5.25}$$

Thereby, the distance l is removed in Eq. (5.25) by using the summation of the T_{cf} and T_{cb}. Here, we show the phase series for T_{cf} and T_{cb} when the person moves forward and backward towards the A_f while running in Fig. 5.39a, b, respectively. They show that the phase values between the two vertical dashed lines experience a large change. For T_{cf}, the phase values first decrease and then increase, while T_{cb} shows the opposite trend. The summation phase values are shown in Fig. 5.39c, in which the large phase changes caused by the random torso movement are offset. In this way, the random effect is eliminated , and we can obtain a relatively clear respiration pattern.

For the period effect which synchronizes with the limb movement, it incurs small and high-frequency fluctuations in the phase values, as shown in Fig. 5.39c. Considering that the limb movement rate is generally higher than the respiration rate during exercise, we apply a third-order Butterworth low-pass filter with f_l as the cutoff frequency to remove the period effect. The phase series after the low-pass filter is depicted in Fig. 5.39d, which shows the periodic pattern of the respiration rhythm.

5.1.3.6 LRC Estimation

For LRC estimation, intuitively, it can be achieved by dividing f_l over f_r and matching the division result to the nearest LRC ratio in Table 5.2. However, this method can lead to inaccurate LRC estimation due to the errors in f_r extraction. There can be cases when f_l and f_r are close to each other in practice. The torso period effect removed by the Butterworth low-pass filter with f_l as the cutoff frequency has residual effects on the respiration pattern due to the existence of the transition band in the low-pass filter's frequency response. Second, if we perform FFT on the filtered respiration signal within a fixed length of the time window,

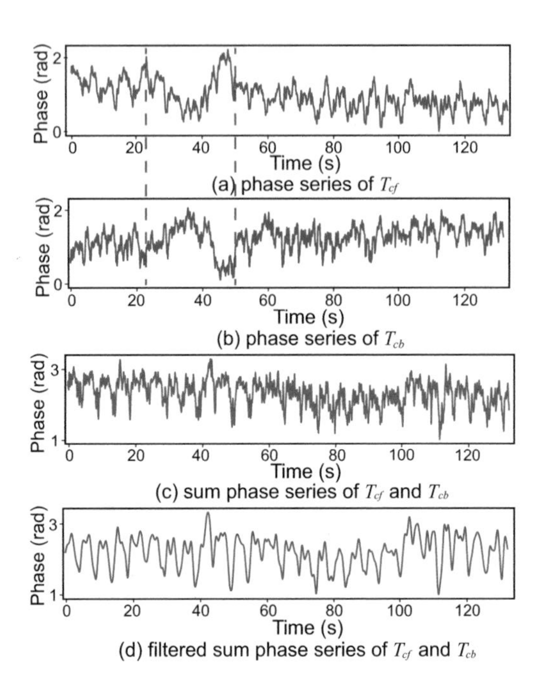

Fig. 5.39 Respiration phase series with effect from torso movements and removal of torso movement effect: (**a**) raw signal phase of the front tag, (**b**) raw signal phase of the back tag, (**c**) summation of signal phases from front and back tags, (**d**) signal phase after the low-pass filter

i.e., 20, for f_r extraction, the resolution of the frequency domain can be relatively low, i.e., $1/20 = 0.05$ Hz. This could bring deviations to f_r estimation within the resolution range.

An example of the FFT series for the filtered respiration signal is shown in Fig. 5.40. We remove the DC component and add a Hamming window on the respiration signal before the FFT operation. The f_r is estimated as the frequency component with the highest peak, where the first yellow dashed line is located at. The second red dashed line corresponds to the real f_r. There is an 0.03 Hz deviation between the estimated and real respiration rates. When calculating the LRC, the nominator (f_l) is usually larger than the denominator (f_r). Consequently, even a small error in the f_r can lead to the wrong LRC estimation. For example, when the real f_l and f_r are 0.6 and 0.24 Hz, respectively, the real LRC ratio should be 5:2. However, if there is only 0.03 Hz error less than the real respiration rate, then the LRC would be wrongly estimated as 3:1.

To estimate the LRC more precisely, we do not extract f_r and calculate the LRC directly. Instead, a correlation-based approach is adopted [12]. The insight of this approach is to obtain the most likely LRC based on the prior knowledge of all the possible LRC ratios and the limb movement rate f_l. In specific, we leverage the given set of LRC ratios r_i for humans during exercise in the LRC list and the estimated f_l, and calculate a set of candidate respiration rates $f_{r_i} = f_l/r_i$. Then, a set of simulated respiration signal is generated with the corresponding f_{r_i} for each of the LRC ratios. The simulated respiration signal is a sinusoidal-like time series, represented as follows:

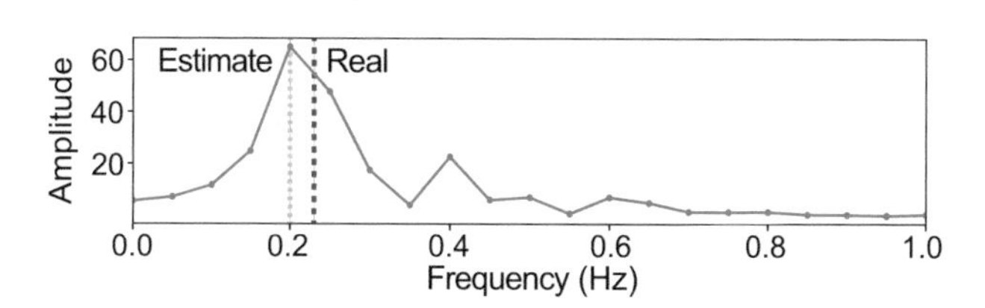

Fig. 5.40 FFT series of the recovered respiration phase values for respiration rate estimation: the yellow dashed line is the estimated respiration rate and the red dashed line is the real respiration rate

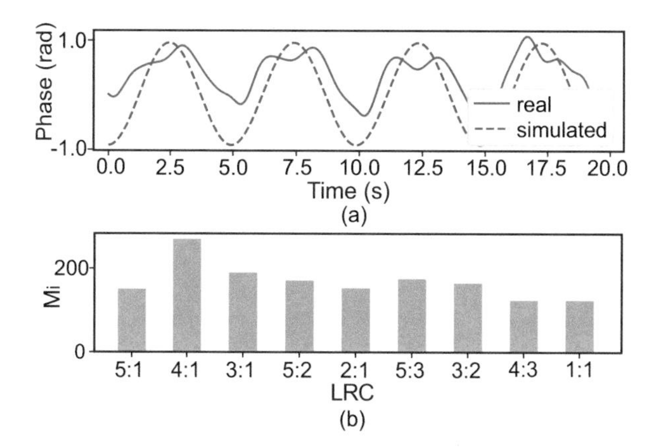

Fig. 5.41 Correlation-based LRC estimation with the example of the real LRC ratio as 4:1. (**a**) Real and simulated respiration signals with 4:1 LRC. (**b**) Max correlation value M_i for different LRC ratios

$$s_i(t) = A * \sin(2\pi f_{r_i} t + \varphi). \tag{5.26}$$

The amplitude of the phase series is normalized to $[-1, 1]$. Thus, A is set as 1, and φ in Eq. (5.26) is 0. Then, we calculate the cross-correlation array between the real phase respiration series φ_{real} and each of the simulated respiration signal $s_i(t)$ for different f_{r_i}. Then, we compare the maximum value $M_i = max\,\{\varphi_{real} \cdot s_i(t)\}$ in the correlation array of different LRCs and choose the LRC with the highest M_i as the final ratio. Figure 5.41a shows an example of the real respiration phase series and the simulated respiration signal with LRC of 4:1. The M_i of different LRC ratios are also depicted in Fig. 5.41b. The maximum M_i is achieved with the candidate LRC of 4:1, which corresponds to the real coupling ratio.

5.1.3.7 Evaluation

Experimental Setup As shown in Fig. 5.42, the ER-Rhythm system is implemented using COTS RFID devices, including the ImpinJ Speedway R420 RFID

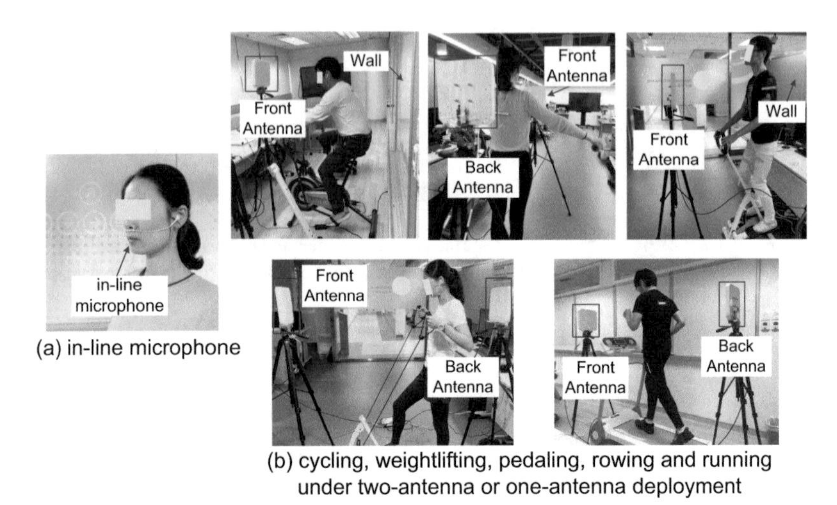

Fig. 5.42 Illustration of the experiment setup: (**a**) setup for ground truth measurement, (**b**) exercise activities and system setup

reader, Laird E9208 antenna, and ImpinJ E41-C tag. The reader works in the 920–925 MHz region, and the reader mode and search mode are set as AutoSetDence and DualTarget, respectively. The reader is connected to a Dell Inspiron 7460 laptop with $i7-7500$U CPU and 8 GB RAM. The RFID measurements are processed with Python 3.0.

We test the system performance in different environments with five exercise activities, as shown in Fig. 5.42b. For cycling and weightlifting, the torso of the exercise could keep relatively stable with small torso movement. For pedaling and rowing, the torso moves along with stepping up-and-down and pulling back-and-forward, in which the periodic effect plays the main role in affecting the respiration pattern. Running brings the highest torso movement effect because the torso will shake fast and randomly with each stride. All five activities involve rhythmic and frequent limb movements.

To measure the ground truth of the limb movement rate, a 3-axis accelerator is worn on the leg or arm. To obtain the ground truth of the respiration cycles, we draw from the fact that people perform nasal breathing with louder breathing sound during exercise than in peacetime [25]. Thereby, we ask all the volunteers to breathe via the nose and stick an in-line microphone under the nose to collect the breathing sound during exercise, as shown in Fig. 5.42a. To guarantee the effectiveness of breathing sound measurement, we find 5 volunteers and collect their breathing sounds with the microphone. They are guided to breathe 50 times with the commands given by another person beside them. Then, we count the number of peaks in the sound waves, from which all the respiration cycles are correctly detected.

In our experiments, 15 volunteers, including 12 males and 3 females, are recruited. They are between 23 and 32 years old ($mean = 26$, $std = 3$), and the

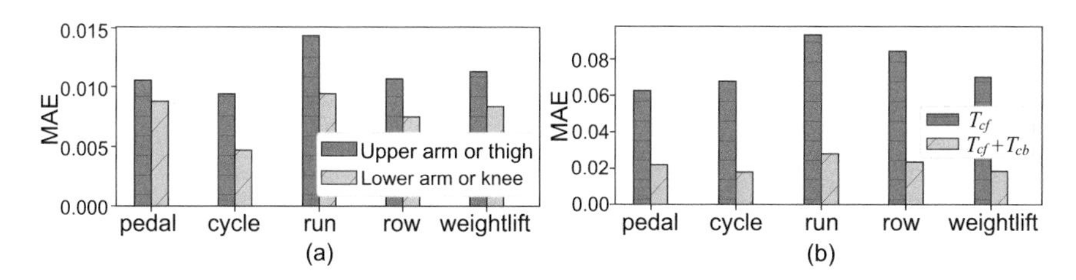

Fig. 5.43 MAEs of estimating limb movement rate and respiration rate. (**a**) Limb movement rate estimation. (**b**) Respiration rate estimation

between 165 and 190 cm tall ($mean = 177$, $std = 6.5$ cm). Some of them are regular exercisers with fitness training at least twice a week, some rarely exercise.

To evaluate the system performance, three metrics are defined: (1) Mean Absolute Error (MAE) of limb movement rate and respiration rate estimation; MAE is calculated as $\frac{1}{n}\sum_{i=1}^{n}|t_i - e_i|$. t_i and e_i are the true and estimated values, respectively. For each time window, the difference between t_i and e_i is calculated, and the MAE is the average error of all the windows. (2) Percentage of Accurate Windows (PAW) of LRC estimation; For a time period, we divide it into several time windows with a fixed length, and the LRC is estimated for each window. Then, we count the number of windows with accurate LRC results over all the windows to get the PAW. For instance, in a 10-min cycling period that involves 30 windows (20s-wide window), the 90% PAW means that 27 windows provide accurate LRC values. We regard the accuracy of LRC is acceptable when PAW is above 90%. (3) Error of LRC estimation; The error of LRC is the difference between the real LRC and the estimated LRC values. For example, the error between the estimated LRC 3:1 and the real LRC 2:1 is 1, the error between the estimated LRC 5:2 and the real LRC 2:1 is 0.5.

Limb Movement Rate Estimation We evaluate the performance of limb movement rate estimation for different exercise activities. For running, pedaling, and cycling, the tag on the leg is used for extracting the limb movement rate. For weightlifting and rowing, the tag on the arm is used. We compare the estimation accuracy of attaching the tags on different limb parts. For the arm, the tag can be put on the upper or lower arm. For the leg, the tag can be put on the thigh or the knee. The results are shown in Fig. 5.43a. The MAEs are all below 0.015 Hz, indicating the estimated rate is quite close to the ground truth. In addition, the MAEs of putting the tags on the lower arm and on the knee are smaller than that on the upper arm and thigh, and the average MAE is around 0.008 Hz. This is because the tags on the lower arm and knee experience longer distance change, so the movement pattern is more obvious.

Respiration Rate Estimation Next, we perform experiments to evaluate the effectiveness of our method for extracting respiration rhythm during exercise and compare with the performance of only using T_{cf}'s phase profile. Although we do

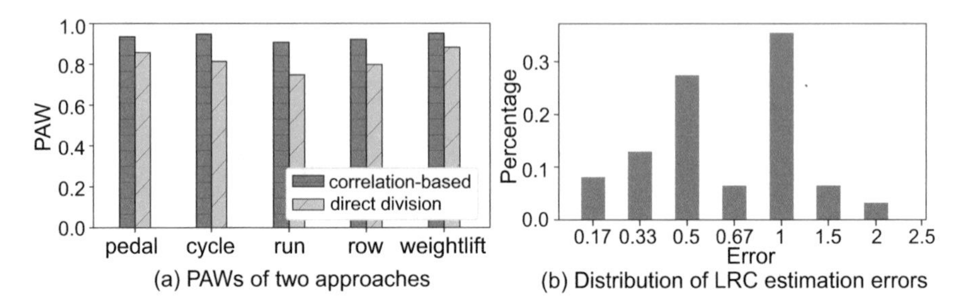

Fig. 5.44 PAWs and errors of LRC estimation: (**a**) comparing our correlation-based method with direct division method, (**b**) distribution of the LRC estimation errors

not calculate the LRC by estimating the respiration rate directly from the summation phase profile, the accuracy of respiration rate estimation is discussed here for comparison. We perform FFT on the summation phase profile of T_{cf} and T_{cb} after chest movement measurement and torso movement effect removal and on the phase profile of T_{cf} after signal pre-processing and low-pass filtering. As depicted in Fig. 5.43b, using only the phase values collected from T_{cf}, the average MAE can be around 0.071 Hz, which is equal to around 4.3 breath per minute. While, with the summation phase values (T_{cf} and T_{cb}), the average MAE is about 0.0225 Hz, corresponding to 1.35 breath per minute. The MAE of respiration rate estimation is reduced by two-thirds.

LRC Estimation The LRC estimation is evaluated with respect to different approaches, distances, and orientations under both two-antenna and one-antenna deployment.

The performance of LRC estimation with different approaches is evaluated under the two-antenna deployment with 1 m distance between the antenna and target human. We first compare the PAW results of directly dividing the limb movement rate over the respiration rate with those of our proposed correlation-based approach, as shown in Fig. 5.44a. The average PAW of the direct division approach is around 81%, which is lower than the average PAW of the correlation-based approach (92%). The main reason is that errors in respiration rate estimation can lead to a large deviation in LRC estimation. While the correlation-based approach narrows the range of possible results by using the potential LRC ratios. We further measure the PAW and main LRC ratios for different activities. The main LRC is the ratio that appears most frequently for the person performing the specific exercise activity. Generally, the PAW results for the running (90.7%) and rowing (92.1%) activities are smaller than those of pedaling (93.5%), cycling (94.8%), and lifting (95.2%). This is because that larger torso movement is involved during running and rowing, which incurs more noises to the respiration pattern extraction. Compared with existing smartphone-based systems [12], whose PAW is around 1–91%, our approach achieves comparable results.

We also show the distribution of the errors between the real and the estimated LRCs for the correlation-based approach in Fig. 5.44b. There are 17 possible error values among all the LRC ratios, while in our experiments, there are 10 error values that appear. The average error of all the samples is 0.74. The frequent error values are 0.5 and 1, and there is no error larger than 2. Thus, the errors in LRC estimation of our system are within a small range. Besides, we pick out all the errors of which the estimated LRC is the neighbor of the real LRC, and the neighbor errors account for 72% of all the errors. This indicates that, even though some errors appear, the estimation LRC ratio is not far from the real one.

5.1.4 UWB Radar-Based Multi-Person Respiration Monitoring

Single-person RM has been realized using WiFi and RFID signals. However, there are few works for multi-person RM, whereas the multi-person scenario is quite common for RM. For example, RM during sleep for an old couple. Therefore, multi-person RM should be properly realized as well. The RFID signal can be used for multi-person RM in a contact-based way by attaching the RFID tags to each of the multiple persons' chests, respectively. Compared with the contact-based manner, multi-person respiration monitoring with contact-free solutions which do not require the attachment of any sensors on the body is more preferred for users' convenience.

Thereby, we attempt to address the above problems for WiFi-based multi-person RM. A potential method is to measure the traveling distance of the signal reflected by different persons' chests. Since multiple persons are located at different positions, the traveling distance of the signal is different. Thereby, we can separate the respiration signal for multiple persons according to the signal traveling distance. However, the resolution of signal traveling distance is quite low since the bandwidth of the WiFi signal is too narrow. The WiFi bandwidth B is 20/40 MHz, so the highest resolution is $(1/B)*c = 7.5$ m where c is $c = 3 \times 10^8$ m/s and $B = 40$ MHz. If two persons are located within 7.5 m with each other, their signal cannot be separated. Some works try to synthesize a wider bandwidth via frequency hopping [36], e.g., 200 MHz, to obtain higher resolution, it is still not enough if two persons are close to each other. Besides, frequency hopping would affect communication among the WiFi AP and its connected devices. In summary, the narrow-band WiFi signal cannot provide precise signal measurements to separate the signal affected by different persons.

To acquire a higher resolution for measuring the signal's traveling distance, the impulse UWB radar with wide bandwidth provides a good solution. Besides, the UWB radar signal is more sensitive to the minute respiration movements. It can also penetrate through the wall which is suitable for indoor short-distance sensing with many obstacles. Meanwhile, there are some commercial and inexpensive products of UWB radar. Hence, after the above attempts and analysis in multi-person RM using the WiFi signal, we decide to make use of the commercial UWB radar to implement a system to realize separate RM for multiple persons. By employing the UWB radar, we can map the respiration state to each corresponding person. Meanwhile, we do not need prior knowledge, e.g., the number of present persons, because we can detect the number of persons and where they are. Thus, in this case study, we employ the spatial-temporal information of the UWB radar signal to estimate the respiration state, i.e., respiration rate and presence of apnea for multiple persons separately.

5.1.4.1 Preliminary of UWB Radar

Our work uses a commercial UWB radar, i.e., Xethru UWB radar, of which the transmitter and receiver (transceiver) are integrated on a single chipset. The UWB radar signal is sent out by the transmitter, gets reflected by the surrounding objects, and gets back to the receiver in a round trip. With the body movements affecting the signal, the time-varying channel impulse response $h(t, \tau)$ of the signal traveling via multiple paths can be expressed as follows [2]:

$$h(t, \tau) = \sum_i \alpha_i \sigma(\tau - \tau_i) + \alpha_r \sigma(\tau - \tau_r(t)),$$

where τ and t refer to the fast-time and slow-time index, respectively. $\sigma(\cdot)$ is the Dirac delta function. The first half term in the above equation denotes the signal reflected by static objects and the second half is related to body movements, *e.g.*, the chest movement during breathing. The respiration incurs periodic movement of the chest, then $\tau_r(t)$ can be represented as a sinusoidal wave as follows:

$$\tau_r(t) = \frac{d_r}{c} = \frac{d_0 + \Delta d \cdot \sin 2\pi f_r t}{c} = \tau_0 + \tau_{\Delta d} \cdot \sin 2\pi f_r t,$$

where τ_0 is the time delay when the person is static, c is the speed of light, Δd is the displacement change of chest movement, and f_r is the respiration rate. The traveling distance of the signal is calculated as $\tau_r(t) * c$. The received signal $r(t, \tau)$ is the convolution of the transmitted impulse $u(t)$ and the channel impulse response, represented as follows:

$$r(m, n) = \sum_i \alpha_i u(n\sigma_\tau - \tau_i) + \alpha_r u(n\sigma_\tau - \tau_r(mT_s)).$$

The received signal is measured in discrete instants along the fast and short time, so τ is represented as $n\sigma_\tau$ and t is converted to mT_s. σ_τ and T_s are the sampling intervals of the fast and short time, respectively. Therefore, the received UWB radar signal is a 3-dimension matrix: fast time n, short time m and signal amplitude $|r(m, n)|$. The fast time and short time correspond to the spatial and temporal dimensions, respectively. The signal matrix is processed to eliminate the components that are reflected by static objects, so that it only contains the dynamic signal which is related to the body movements. Here, the sampling rate of the used UWB radar is 17 Hz. The human respiration rate is usually lower than 0.6 Hz. Thus, the sampling rate meets the requirement of the Nyquist sampling theorem. The Pulse-Repetition Frequency (PRF) of the UWB radar is 15.875 MHz. The maximum unambiguous range $R_{max} = c/(2 * PRF)$ is around 10 m. Then, the maximum detected distance is set as 10 m. The UWB radar reports the signal of all traveling distance points with the resolution (I_u) of 5.14 cm.

5.1.4.2 Respiration Pattern in the UWB Radar Signal

With the presence of multiple persons at different locations, the signal matrix will show the respiration patterns along the short-time axis for each person in different sets of fast-time indexes, i.e., distance points. An example of the signal matrix when two persons breathe at the same time is depicted in Fig. 5.45a. Each time series shows the signal amplitude during 60 s at each distance point. The two persons, P_1 and P_2, are located at around 1 and 1.5 m away from the UWB radar transceiver, respectively. The respiration activities of the two persons incur the signal amplitude of several consecutive rows to change periodically, especially for the person located closer to the transceiver. It also shows that the respiration activity of P_1 has a larger amplitude change than that of P_2.

We further show the multiple time series revealing the respiration pattern of P_2 in Fig. 5.45b. There are many noises in each time series due to the multipath effect caused by the presence of P_1, and the respiration pattern is not clearly periodic. For respiration state extraction, existing works mainly select one time series for analysis. For example, if we choose the time series with the highest standard deviation, the noises and fake peaks in the time series can lead to inaccurate results of the respiration state. What's more, traditional methods using time series analysis for respiration state estimation, e.g., frequency-domain analysis [33] or peak detection [17], may not be effective in face of the imperfect respiration patterns. As a comparison reference, we choose one time series with the highest standard deviation from the signal matrix and perform Fast Fourier Transformation on it to extract the respiration rate. The estimation error can reach around 1.5–2 bpm, as illustrated in

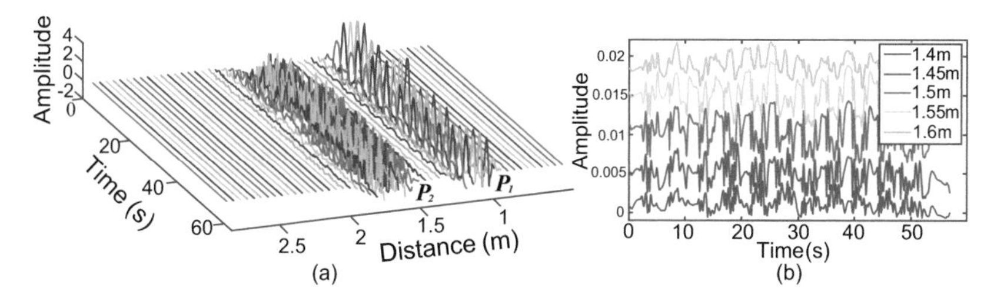

Fig. 5.45 (**a**) Illustration of the signal matrix while respiration of two persons, (**b**) Raw time series revealing the respiration pattern of P_2

the evaluation section (Fig. 5.53), which is quite large compared with the results of state-of-art systems (0.5–1 bpm). Therefore, we need to find another method to get rid of those noises for precise respiration state estimation, which will be introduced in the next section.

5.1.4.3 Presence Detection

The first task is to detect the presence of all the persons. Here, we adopt a threshold-based approach for presence detection. Previous methods usually find a fixed threshold for single-person detection, which is not suitable for our multi-person scenario. In Fig. 5.45a, we show that the farther the person is away from the transceiver, the lower the signal amplitude change will be. In this way, a fixed threshold would not be effective to detect the presence of all the persons located at different locations. Hereby, we propose to assign a set of dynamic thresholds for each distance point in the signal matrix. From the power loss model of the RF signal represented in the Friis equation, the power of the received signal can be expressed as follows:

$$P_r = \frac{P_r G_t G_r \lambda^2}{(4\pi)^2 d^2}.$$

It reveals that the signal strength falls off proportionally to the square of the signal traveling distance d. If the transmitting power P_t, antenna gain G_t, G_r, and the signal wavelength λ remain unchanged, the standard deviation (SD) of P_r will only change with the distance.

$$SD(P_r) \sim 1/d^2. \tag{5.27}$$

To verify Eq. (5.27) in practice, we measure and calculate the standard deviation of the signal amplitude when the person breathes normally along different distance points to the transceiver. The results in Fig. 5.46 show that the measured SD approximate to the theoretical SD well. Thereby, we follow Eq. (5.27) to set a series of thresholds according to a selected baseline threshold that changes dynamically.

First, we calculate the standard deviation of the signal amplitude at each fast-time index, i.e., distance point. Second, the highest deviation and its adjacent two points are used to get the average value of them as p_m. The distance point of the highest deviation is d_0. Then the baseline threshold σ_0 is settled as $\frac{(p_m - n)}{(2d_0+1)} + n$. Here, n is the noise value which is the mean of the deviation vector. With the increment of d_0, σ_0 decreases inversely. Then, the threshold σ_i at distance d_i can be obtained based on the baseline as follows:

$$\frac{SD(P_{d_0})}{SD(P_{d_i})} \sim \frac{1/d_0^2}{1/d_i^2} = \frac{d_i^2}{d_0^2} = k, \text{ and } \sigma_i = \frac{1}{k} \cdot \sigma_0.$$

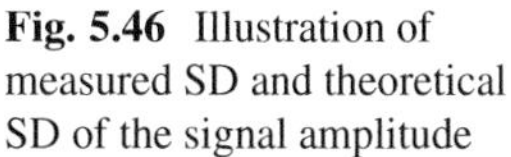

Fig. 5.46 Illustration of measured SD and theoretical SD of the signal amplitude

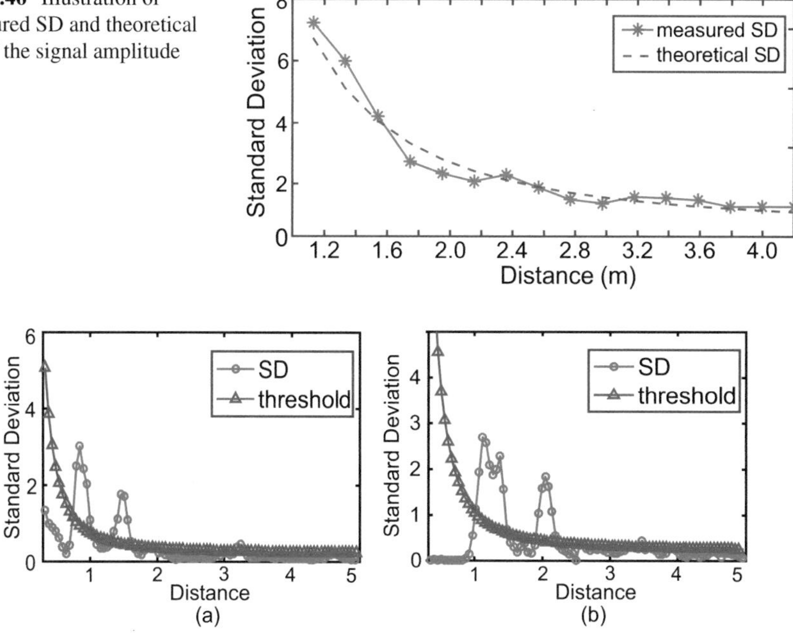

Fig. 5.47 Threshold set and measured SD: (**a**) two persons breathe together, (**b**) one person moves backward, one person breathes

By comparing the SD values of each distance point with the set of thresholds σ, the consecutive time series whose deviations are above σ are selected out to form a sub-matrix. In Fig. 5.47a, the real SD values and the calculated threshold set are shown when two persons breathe together, the first two peaks indicate that there are two persons in the environment. The minor peak at around 3.2 m is caused by the outliers, which need to be removed.

To remove the signal outliers for accurate presence detection, we only keep the sub-matrices that involve 4–8 distance points since the chest is about 15–45 cm wide for most common people. Then, the central distance point of the multiple rows is regarded as the person's distance to the radar. If the number of consecutive rows is smaller than 4, they will be discarded as signal outliers. If the number is larger than 8, the corresponding person will be regarded as performing the large-scale movement. This is because the large-scale movement could affect the radar signal in a wider and longer range. In Fig. 5.47b, the SD values when the person at around 1 m moves one step backward is shown. The number of SD values above the thresholds of this person is 11. Once the large-scale movement is detected, RM will be stopped temporally until the person returns to the quasi-static state. In case that multiple persons might have the same distance to one transceiver, one more transceiver can be employed to get the absolute location of each person in the 2D space. Then, we only keep the sub-matrix from the transceiver, which is closer to the person, for estimating the respiration state.

5.1.4.4 Respiration State Estimation

After presence detection, the second goal is to estimate the respiration state, including the respiration rate and the presence of apnea, from the signal sub-matrix of each person. Previous works extract the respiration state by analyzing a single time series of the RF signal. However, in the evaluation section, the comparison results in Fig. 5.53 show that the performance on respiration rate estimation is imperfect using the traditional time series analysis approach. This is mainly due to the reason that the presence of multiple persons incurs many noises in the signal time series. Therefore, it is not an effective way to apply previous time series analysis methods to extract the respiration state. Nevertheless, through our observation, we find that by converting each signal sub-matrix within a certain period (e.g., 30 s) into an RGB image, the image can show a clear and periodic respiration pattern. For instance, we show the images of the sub-matrices with two persons and three persons breathing concurrently in Fig. 5.48a, d, respectively. Here, the separate images of the sub-matrices for each person are concatenated together with black blocks as intervals for easy viewing. Each sub-matrix first goes through a normalization operation before being transformed into the image. The change of light and dark areas clearly reveals the breathing and non-breathing exchange.

Edge Detection To extract the respiration state from the image, we first detect the edges between the breathing and non-breathing intervals. To make the edge easier to be detected, we first apply the L0-norm gradient minimization on the image for edge-preserving smoothing. It helps to remove noises and unimportant details in the image and make the edge more salient for detection. The output image S is acquired after solving the following objective function:

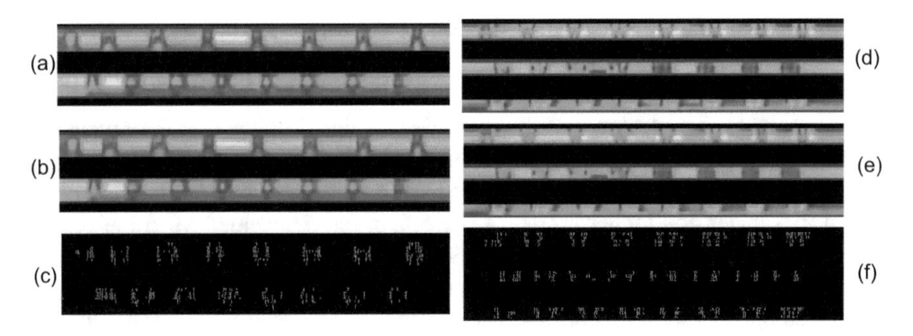

Fig. 5.48 Original image transformed from sub-matrix: (**a**) two persons, (**d**) three persons. Image after L0-norm gradient minimization: (**b**) two persons, (**e**) three persons. Image after vertical Sobel edge detection: (**c**) two persons, (**f**) three persons

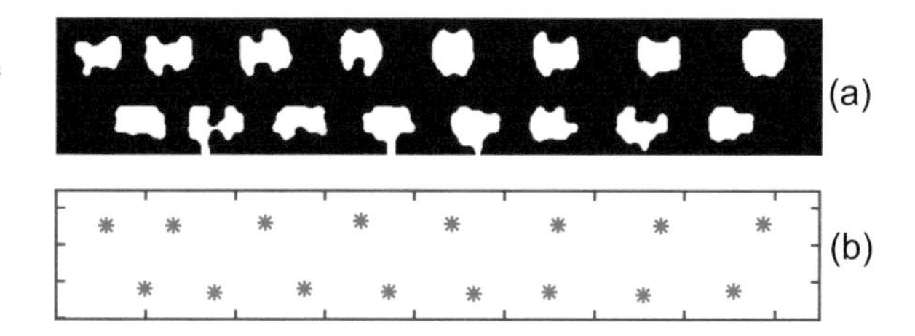

Fig. 5.49 Two-person respiration pattern: (**a**) Image after performing dilation and erosion on the edge-detected image, (**b**) blob detection result

$$\min_{S}\{\sum_{p}(S_p - I_p)^2 + \lambda \cdot C(s)\}, s.t., C(s) = \kappa,$$

$$C(s) = no.\{p|\left|\partial_x S_p\right| + \left|\partial_y S_p\right| \neq 0\}.$$

I is the input image, and p is the pixel. The parameters λ and κ, which control the degree and rate of smoothing, are set at 2×10^{-2} and 10 after testing with a number of trails. Next, the vertical Sobel edge detection which can detect the edges only in the vertical direction, is applied on the smoothed image twice. The first Sobel operator is used for deciding the initial threshold for edge detection. To enhance the obscure edges, the final threshold is set at 5% less than the initial one. Then, the second Sobel operator is applied, and the edges are extracted. The images after edge-preserving smoothing and edge detection are shown in Fig. 5.48b–c and e–f, respectively. After edge detection, there are groups of vertical edges showing the repetitive patterns of the breathing cycle.

Respiration Pattern Extraction With the detected edges, we need to extract the respiration rate and detect the apnea. First, we perform image dilation on the vertical edges so that each group of the spaced edges can be expanded into a closed area. Then, image erosion is applied to the dilated image to shrink the closed areas to capture a clear respiration pattern. There can be some tiny areas in the image as noises. The Gaussian low-pass filter is leveraged to denoise the image. As examples, the images for two-person and three-person respiration after dilation, erosion, and filtering are shown in Figs. 5.49a and 5.50a, respectively. The presence of the white closed areas refers to the alternation of breathing cycles.

To estimate the respiration rate, we treat the closed areas as blobs and perform blob detection to count the closed areas as the number of breathing cycles. In case that some tiny areas are not cleaned out after filtering, we first remove the blobs whose area is smaller than the quartile of all the blobs. Then, each closed area will be labeled as one point, which is the center of the area. The results after implementing blob detection are given in Figs. 5.49b and 5.50b, respectively. Here, each start point represents a respiration cycle. Thus, the respiration rate can be calculated by counting the number of points P in a certain time T (in seconds), and the respiration rate $Rate$ can be calculated as $Rate = 60 * P/T$.

However, there can be some fake blobs which do not refer to the real breathing cycle in the images, as a result of the presence of some large noisy areas after image

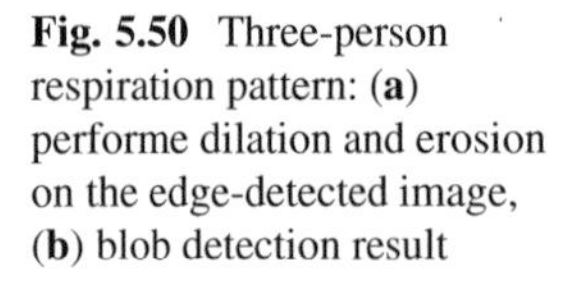

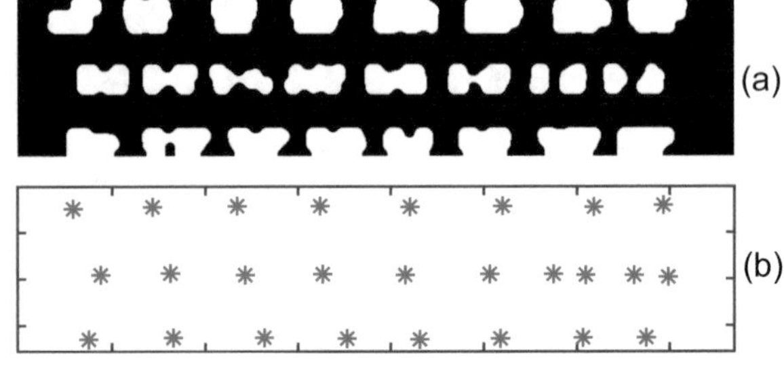

Fig. 5.50 Three-person respiration pattern: (**a**) performe dilation and erosion on the edge-detected image, (**b**) blob detection result

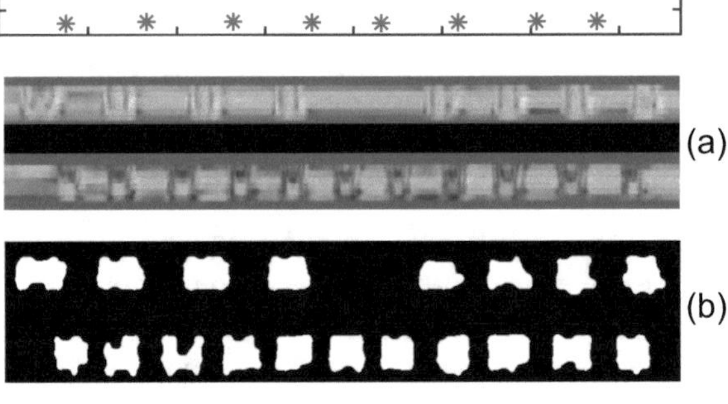

Fig. 5.51 Apnea detection: (**a**) original RGB image, (**b**) after blob detection

dilation and erosion. For example, in Fig. 5.50, the last four closed areas in the second row actually refer to two respiration cycles. The two blobs belonging to one breathing cycle are quite near to each other. To remove the fake blobs, we integrate the two blobs of which their centroids are within a given range and make the mean value of the two centers as the new blob center. Since the human respiration rate is commonly in the range of (10, 40 bpm), a single breathing cycle would take up at least 1.5 s. Here, we transform the 30 s duration signal matrix into an image with the size of $M * N$ ($M = 650$, $N = 850$). If the horizontal distance of the centroid of adjacent blobs is smaller than the threshold α_1 ($\alpha_1 = 850/30 * 1.5 = 42$ pixels), then the two blobs will be integrated into one blob.

Finally, the disappearance of blobs is monitored to detect the apnea syndrome. As an example, we show the original RGB image and the extracted blob image when one of the two persons stops breathing, i.e., holding the breath for about 7 s in Fig. 5.51. The long interval between the 4th and 5th blobs in the first row of Fig. 5.51b is caused by the stop of respiration. We leverage this observation to detect the apnea. First, we choose a threshold α_2 for comparing the horizontal length of the interval between two neighbor blobs. The minimum normal respiration rate is 10 bpm, meaning that the longest breathing cycle lasts for around 5 s. Meanwhile, the presence of apnea syndrome usually lasts for 10–30 s. Thus, we set the threshold α_2 for apnea detection as 5 s, which corresponds to 140 pixels in the image. Then, we can demonstrate that there appears the apnea if the horizontal distance of two adjacent blobs is longer than 140 pixels.

5.1.4.5 Evaluation

We use a commercial UWB radar for experiments. The UWB radar chip is around US$50 and designed to comply with the FCC regulations and would not post risks on human health. The chipset is tied on the tripod with 1.5 m high. The detailed parameters of the UWB radar are given in Table 5.3. The chipset is connected to a laptop which collects the received signal for processing. We recruit 10 volunteers, including 7 males and 3 females, and divide them into 2 groups for separate experiments. The UWB radar signal is measured with different numbers of persons, i.e., 1 to 5 persons for each group. For the first group, it involves 3 males and 2 females, and the average height and chest breadth among them are 173 and 34 cm, respectively. The second group consists of 4 males and 1 female with the mean height and chest breadth of 175 and 36 cm, respectively.

The UWB radar signal is collected under different scenarios (standing, sitting, lying) and postures (front or side) for the investigation of their effects on the performance. For the lying scenario, people lie abreast along each other and cover themselves with a thin blanket to simulate sleeping. They are located at different distances to the radar transceiver, thus only one chipset is employed. For the sitting and standing scenarios, in case people at different locations can have the same distance to the same transceiver, two chipsets are used to eliminate the distance ambiguity in the 2D space.

Evaluation Metrics To evaluate the performance of presence detection, we define the accuracy of presence detection as the ratio between the number of correctly detected samples and the number of all the samples. The number of missed or extra detected persons is calculated as the error of presence detection.

To evaluate the performance of respiration rate estimation, we define the Mean Absolute Error (MAE) between the estimated respiration rate and the real one. MAE is calculated as $\text{MAE} = \frac{1}{n}\sum_{i=1}^{n}|r_i - \bar{r_i}|$. r_i and $\bar{r_i}$ are the real and estimated respiration rates, respectively. Here, n is the number of samples. The respiration rate is measured in breath per minute (bpm). During experiments, each person wears a chest belt with a built-in 3-axis accelerometer to record the real respiration state which is regarded as the ground truth r_i.

To evaluate the performance of apnea detection, we calculate the percentage of missed apnea (MA) as the number of missed apnea cases over all the real apnea cases. In addition, the percentage of false apnea (FA) is measured from the number

Table 5.3 Parameters of the leveraged UWB radar

Sampling rate	Pulse-repetition frequency
17 Hz	15.875 MHz
Effective range	Distance resolution
10 m	5.7 cm
Low frequency band	High frequency band
6.0–8.5 GHz	7.25–10.2 GHz

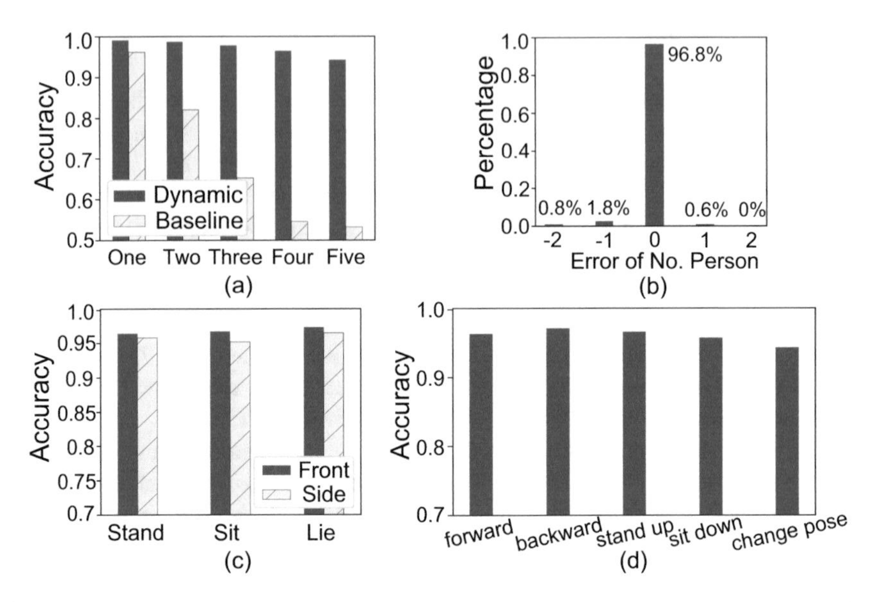

Fig. 5.52 Performance of presence detection: (**a**) accuracy of different numbers of persons, (**b**) distribution of errors of the number of persons, (**c**) accuracy under different scenarios and postures. Large-scale movement detection: (**d**) accuracy of detecting different large-scale movements

of false detected apnea over all the respiration cases without apnea. MA and FA are represented as follows:

$$\text{MA} = No.(missed\ apnea)/No.(real\ apnea) \times 100\%,$$

$$\text{FA} = No.(false\ apnea)/No.(No\ apnea) \times 100\%.$$

Performance of Presence Detection We first evaluate our method of applying the set of dynamic thresholds to detect the presence of multiple persons. Here, different numbers of persons are asked to stay in the vicinity of the UWB radar transceiver. The involved persons are all quasi-static and breathing normally. The accuracy of presence detection with different numbers of persons and the distribution of detection errors on the number of persons are shown in Fig. 5.52a, b. We also compare the accuracy of applying dynamic thresholds with that of a fixed baseline threshold in Fig. 5.52a. It shows that our method can achieve more accurate results for presence detection, especially for the multi-person scenario. The average accuracy on presence detection using the dynamic thresholds is 95.7%, while the average accuracy with the fixed baseline threshold is only 69.4%. With the increasing of present persons, the detection accuracy drops gradually, and it is more likely to miss the detection of 1 or 2 persons because the front person can block off the person behind.

In practice, the missed detection of a person at one moment can be calibrated later if the person moves. Thereby, long-term monitoring would provide more accurate results for presence detection. We also consider the effects of different scenarios and postures on presence detection, of which the results are illustrated in Fig. 5.52c. The performance of presence detection when people face front to the UWB radar transceiver is slightly better than that of on the side. Meanwhile, for the standing, sitting, and lying scenarios, the presence detection accuracy similarly fluctuates around 94 to 96%.

We also evaluate the detection of large-scale movements compared to the quasi-static state during breathing, like moving around or changing postures. If large-scale movements are detected, the corresponding sub-matrix will be omitted for respiration state estimation. The accuracies for detecting the large-scale movements, including moving forward and backward, standing up and down, changing lying postures, are shown in Fig. 5.52d. It shows the average accuracy of large-scale movements detection is around 95%. While, changing postures between the front and the side has the lowest accuracy due to the reason that people tend to change posture in place, resulting in fewer location changes.

Performance of Respiration Rate Estimation For respiration rate estimation, we first show the MAEs for different numbers of persons. The average MAEs for 1–5 persons in the two groups are illustrated in Fig. 5.53 when they are sitting while breathing. Generally, the average MAEs go up with the increasing number of present persons from 0.185 bpm for 1 person, 0.282 bpm for 2 persons, 0.342 bpm for 3 persons, 0.413 bpm for 4 persons and 0.558 bpm for 5 persons. The worst result is that the MAE reaches 0.58 bpm. The accuracy degradation of the respiration rate estimation when more persons are at present is mainly due to the adverse multipath effect caused by the movements of others. We also compare the estimation accuracy by using one time series from the signal sub-matrix with our image-based method in Fig. 5.53. The average MAE can reach as high as about 2 bpm for five persons. The comparison results reflect the effectiveness of our method.

Performance of Apnea Detection Since the respiration apnea syndrome usually happens during sleep, we simulate the apnea by asking the involved persons to lie down and hold their breath for a few seconds ($\geq$ 6 s). Then, we detect the presence of apnea with 1–5 persons, and one of them will act as the person who suffers from apnea. The percentage of missed and false detected apnea is shown in Fig. 5.54. The percentage of missed apnea is between 3 and 5%, and the percentage of false detected apnea is in the range between 4 and 7%. With more persons at present, the accuracies of apnea would also drop down gradually.

It also shows that the percentage of false detected apnea is slightly higher than that of missed detected apnea. For apnea detection, it would be acceptable for the users that the apnea is sometimes reported when there is no apnea. While, if the system misses the real apnea, which means that the MA is much higher, then it may result in a serious consequence for the person who suffers from the apnea.

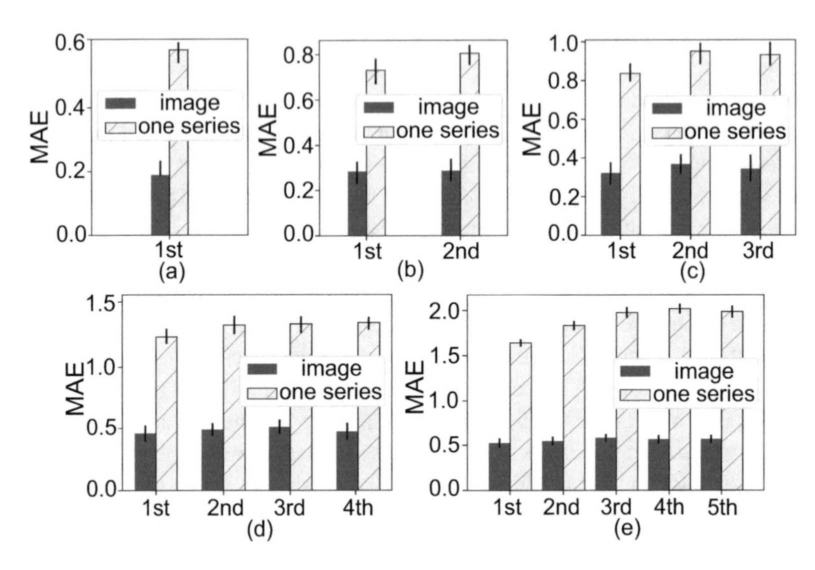

Fig. 5.53 MAEs of respiration rate estimation and accuracy comparison: (**a**) 1 person, (**b**) 2 persons, (**c**) 3 persons, (**d**) 4 persons, (**e**) 5 persons

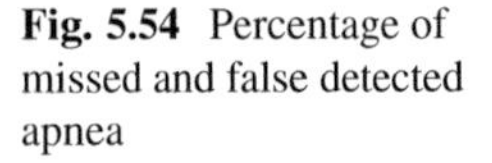

Fig. 5.54 Percentage of missed and false detected apnea

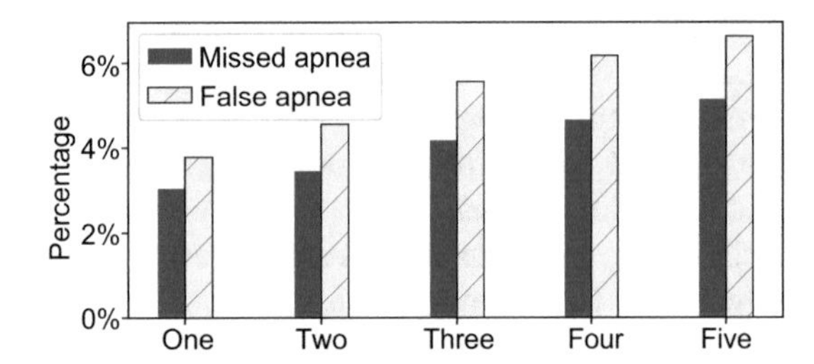

5.2 Human and Object Indoor Localization

Indoor localization is the process of obtaining the position of a person or an object using sensory information of wireless signals in an indoor setting or environment. It has increasingly drawn interests from both academia and industry since it has real significance and wide applications, such as indoor navigation and object tracking. There are various available wireless technologies for the design of an indoor localization system. In this section, we will showcase our studies in Bluetooth-based and RFID-based indoor localization. First, we utilize the received signal strength indicator (RSSI) of Bluetooth for localizing the Bluetooth device on the sensing target. As we have introduced in Sect. 4.2.1.1 that the signal strength is an unreliable indicator for estimating the distance using the Friis equation. Therefore, we present a data-driven approach and design a neural network to enhance the localization accuracy. Second, we introduce a mobile RFID system for localizing objects attached with RFID tags in a large warehouse. Since the data-driven approach requires the cumbersome collection of a large amount of data, we adopt a model-based approach to derive the object location in three dimensions with labor-intensive data collection.

5.2.1 Bluetooth-Based Human Indoor Localization

Bluetooth Low Energy (BLE) technology has the advantages of longer communication range and extremely low power consumption, meanwhile it can achieve comparable localization accuracy with other existing wireless technologies. In existing Bluetooth 5.0 standard, RSSI and other auxiliary information are measured, which means BLE indoor localization is a range-based localization method. It calculates the position of a point by distance which is predicted based on measured RSSI, using the geometry of circles or spheres. Conventionally, trilateration is one of the representative range-based localization algorithm, in which conventional radio propagation model is used to estimate the distance between device and surrounding anchor points (nodes with known fixed positions) based on collected RSSI measurements. With the estimated distances to three anchor points nearby, device's location can be obtained by solving the relative geometric relationship among triangles. Least square and weighted centroid are two common methods solving the trilateration problem.

Least square method first defines a set of equations that describe the distance relationship between the device we try to locate and the anchor point with fixed position, which are subsequently solved using least square minimization formula. On the contrary, weighted centroid measure tackle this problem in a more direct way that it first attempts to get the intersections of every two of three circles, then takes the average of them to represent the estimated position. Theoretically, it seems quite natural for these methods to obtain good localization results.

However, in practice, their performances are far less than satisfactory to achieve high localization accuracy. The reason for this is twofold: First of all, radio environment is changing fast since it is interfered, obstructed and attenuated during propagation; Secondly, due to the sensitivity and stability issues of hardware devices, the collected measurements is not always 100% reliable. Moreover, estimating the distance between device and anchor point is critical in range-based localization, but traditional propagation models fail to make accurate prediction as they are theoretical models whose parameters are empirically gained such that they are unable to characterize the complicated non-linear decay of wireless signal. In addition, system error aggregates as different parts are joined together, which can lead to instability and inaccuracy of the system's performance.

To address aforementioned shortcomings, we propose a pure deep learning based model to make the distance estimation, which is consisted of a long short-term memory (LSTM) network followed by a multi-layer fully-connected neural network. Benefiting from the strong capability of non-linear feature representation of deep neural networks, LSTM-based estimator can better model attenuation of wireless signal and improve the distance prediction accuracy. Besides, a bunch of self-adaptive mechanisms are employed to counteract the side effects brought by incoming undependable measurements.

5.2.1.1 LSTM-Based Self-Adaptive Localization

We present details on localization process, which can be divided into two parts: Phase I: Offline Training on Distance Estimator and Phase II: Online Localization and Tracking.

Phase I: Offline Training on Distance Estimator
At present, Bluetooth indoor localization is a range-based localization algorithm solely relying on measured RSSI. Distance estimator attempts to transform the collected measurements to the distance between moving user tag and mounted Bluetooth gateway (anchor point with fixed position), indicated by d_{tag-gw}^{prop}, and accurate distance prediction is the prerequisite to make precise positioning. In Bluetooth communication, the distance calculated by traditional propagation model can be simply expressed as [42]:

$$d_{tag-gw}^{prop} = 10^{\frac{|RSSI|-A}{10\times n}}, \tag{5.28}$$

with A and n two empirical parameters. A represents the measured RSSI when the distance between the user tag and the gateway is exactly 1 m, and n is the environmental attenuation factor. However, due to the extreme complexity of radio environment, it fails to precisely model the nonlinearity of the propagation characteristics of wireless signal.

To address the aforementioned limitations, rather than using propagation model, we design a completely new distance estimator that is fully based on deep learning models. The basic structure of this model is illustrated on the lower-right of Fig. 5.55. We first use a long short-term memory (LSTM) network to address the time dependency of sequential measured RSSI data under the assumption that for a very short period of time Δt, say $\Delta t < 1$s, people to be positioned can be regarded as remaining stationary.

$$\begin{cases} \mathbf{i}_t = \sigma(\mathbf{w}_i \mathbf{x}_t + \mathbf{u}_i \mathbf{h}_{t-1} + \mathbf{b}_i), \\ \mathbf{f}_t = \sigma(\mathbf{w}_f \mathbf{x}_t + \mathbf{u}_f \mathbf{h}_{t-1} + \mathbf{b}_f), \\ \mathbf{o}_t = \sigma(\mathbf{w}_o \mathbf{x}_t + \mathbf{u}_o \mathbf{h}_{t-1} + \mathbf{b}_o), \\ \tilde{\mathbf{c}}_t = \tanh(\mathbf{w}_{\tilde{c}} \mathbf{x}_t + \mathbf{u}_{\tilde{c}} \mathbf{h}_{t-1} + \mathbf{b}_{\tilde{c}}), \\ \mathbf{c}_t = \tilde{\mathbf{c}}_t \otimes \mathbf{i}_t + \mathbf{c}_{t-1} \otimes \mathbf{f}_t, \\ \mathbf{h}_t = \mathbf{o}_t \otimes \tanh(\mathbf{c}_t), \end{cases} \tag{5.29}$$

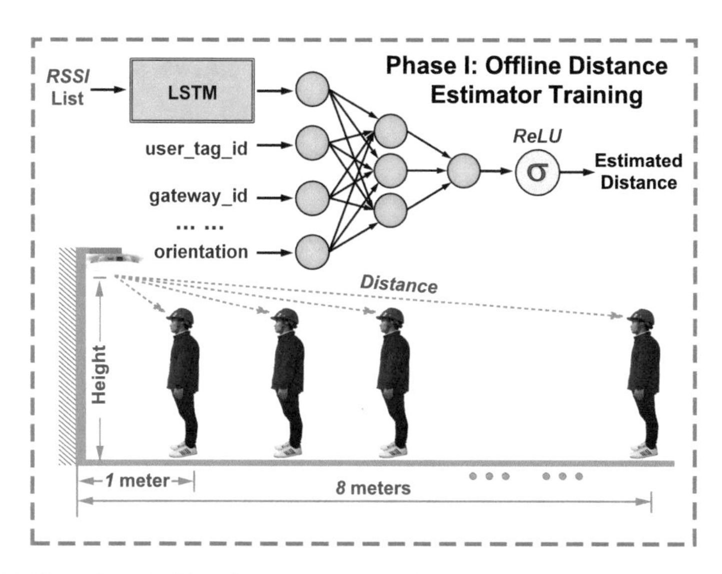

Fig. 5.55 Illustration of offline distance estimator training. (**a**) LSTM based distance estimator. (**b**) Collect the ground truth every 1 m

where **i**, **f**, **o** are used to denote *input*, *forget* and *output gates*, respectively. $\mathbf{x}_t$ that equals $rssi_t$ is the input to LSTM at time t $(1<t<T)$ and $\mathbf{h}_t$ is the corresponding hidden state, which can be taken as the output of LSTM at that moment. $\otimes$ represents the element-wise multiplication operator, and $\sigma(\cdot)$ and $\tanh(\cdot)$ are non-linear *sigmoid* and *tanh* activation functions (i.e., $\sigma(x) = \frac{1}{1+e^{-x}}$ and $\tanh(x) = \frac{1-e^{-2x}}{1+e^{-2x}}$).

Then, the latest element of LSTM output $\mathbf{h}_T$ is concatenated with tag_id, $gateway\_id$ and other features, i.e.,

$$\mathbf{y}_{\text{CAT}} = [\, \mathbf{h}_T,\ tag\_id,\ gateway\_id,\ \ldots]. \tag{5.30}$$

Then, $\mathbf{y}_{\text{CAT}}$ is further processed by a fully-connected multi-layer perceptron neural network $\mathbf{MLP}(\cdot)$ followed by a non-linear *sigmoid* activation function $\sigma(\cdot)$ as

$$d^{DL}_{tag\text{–}gw} = \sigma\big(\mathbf{MLP}(\mathbf{y}_{\text{CAT}})\big). \tag{5.31}$$

After setting up the model, we have to do abundant measurement work to collect ground truth labels for training the LSTM-based distance estimator. More concretely, we measure RSSIs from different directions starting from the point whose distance to the gateway $d_{tag\text{–}gw}$ is 1 m sharp, then repeat the measurements every 1 m until move to the point 8 meters away from the gateway. It is unnecessary to measure RSSIs as the distance between the gateway and the user tag is longer than 8 meters, because the RSSI is indistinguishable when $d_{tag\text{–}gw} > 8\,m$. With the

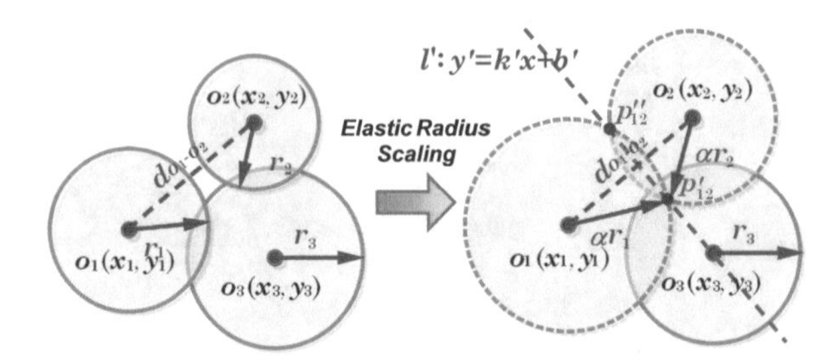

Fig. 5.56 Illustration of elastic radius intersecting

well-trained distance estimator, the next step is to perform online localization and tracking.

Phase II: Online Localization and Tracking

In localization algorithm, trilateration is the process of determining the absolute or relative location of a device using the geometry of three circles. Online localization and tracking module generally follows the process of classic trilateration method, but shows some differences. Simple trilateration is a pure RSSI-based localization algorithm, whose obtained result is vulnerable to high fluctuation of measured RSSIs. To enhance the localization accuracy and stability against unreliable RSSI measurements, in our proposed localization algorithm, a series of self-adaptive mechanisms are adopted, including elastic radius intersecting, multiple weighted-centroid localization and self-adaptive Kalman tracking, which will be justified in details below.

In conventional trilateration, it is a common occurrence that any two of three circles with centers $\o_1$ at (a_1, b_1) and o_2 at (a_2, b_2) may have no intersection since the estimated distances between the device (user tag, say **tag**1) and the anchor points (Bluetooth gateways, say **gw**1 and **gw**2) are shorter than their actual values. That is, as illustrated in Fig. 5.56, $r_1 + r_2 < d_{o_1-o_2}$, where $d_{o_1-o_2}$ is the distance between o_1 and o_2, and $r_1 = d^{DL}_{tag1-gw1}$ and $r_2 = d^{DL}_{tag1-gw2}$ are the corresponding radii of **circle**1 (c_1) and **circle**2 (c_2), respectively. It may lead to localization failure if two circles are not intersected.

To address this problem, we intuitively take a simple but effective approach, called "elastic radius intersecting". It scales up the radii of both circles by multiplying them with a scaling factor α so as to ensure the two circles can be intersected, i.e., $\alpha r_1 + \alpha r_2 \geq d_{o_1-o_2}$.

It is worth noting that when picking the scaling factor $\alpha \in \{\alpha_1, \ldots, \alpha_M\}$ ($\alpha_1 < \ldots < \alpha_M$ is an arithmetic progression), we first set $\alpha = \alpha_1$ and try to find the intersection, and if it fails to make it, then α is enlarged to α_2 and continue using progressive method until an intersection is obtained or α reaches its upper limit (i.e., $\alpha \geq \alpha_M$). Empirically, we set $M = 10$ and $\alpha_M = 1.5$, which aims to restrict the scaled radius within a reasonable range.

After scaling, there is much higher possibility that two circles can get intersected. As illustrated in Fig. 5.56, assume there are two intersections between **circle1** and **circle2**, i.e., $p'_{12} : (x_{p'_{12}}, y_{p'_{12}})$ and $p''_{12} : (x_{p''_{12}}, y_{p''_{12}})$, we can get the line passing through them: $l' : y = k'x + b'$, where $k' = \frac{a_1-a_2}{b_2-b_1}$ and $b' = \frac{(\alpha^2 r_1^2 - \alpha^2 r_2^2)-(a_1^2-a_2^2)-(b_1^2-b_2^2)}{2(b_2-b_1)}$. Subsequently, according to trigonometric calculation, we have

$$(\alpha r_1)^2 = (x_{p'_{12}} - a_1)^2 + (y'_{p_{12}} - b_1)^2, \tag{5.32}$$

in which $y'_{p_{12}} = k'x'_{p_{12}} + b'$. By solving Eq. (5.32), we can get

$$x'_{p_{12}}, x_{p''_{12}} = \frac{-B \pm \sqrt{B^2 - 4AC}}{2A}, \tag{5.33}$$

where $A = (k')^2 + 1$, $B = -2(a_1 + k'b_1 - k'b')$ and $C = a_1^2 + b_1^2 - 2b'b_1 + b'^2 - \alpha^2 r_1^2$. Then, $y'_{p_{12}}$ and $y_{p''_{12}}$ can be calculated accordingly. In practice, we only keep the intersection which is closer to **circle3**, i.e., $p'_{12} : (x_{p'_{12}}, y_{p'_{12}})$ in the illustration, while the farther one $p''_{12} : (x_{p''_{12}}, y_{p''_{12}})$ will be discarded. For simplicity, we use $p_{12} : (x_{p_{12}}, y_{p_{12}}) = p'_{12} : (x_{p'_{12}}, y_{p'_{12}})$ to denote the unique intersection of interest.

Usually, in trilateration, we use the centroid to represent the approximate estimated position of the device to be located, which is computed as the mean of three intersections if any two circles of three intersect with one another. For example, let's assume p_{12}, p_{23} and p_{31} are the three obtained intersections, and traditionally, the centroid of these three points $c_{123} : (x_{c_{123}}, y_{c_{123}})$ can be calculated as the arithmetic average: $x_{c_{123}} = \frac{x_{p_{12}}+x_{p_{23}}+x_{p_{31}}}{3}$ and $y_{c_{123}} = \frac{y_{p_{12}}+y_{p_{23}}+y_{p_{31}}}{3}$. Instead of simple average, a more convincing way is to use the weighted average of three centroids to represent the estimated position as follows:

$$\begin{cases} x_{c_{123}} = \sum_{i,j \in \{1,2,3\}} w_{p_{ij}} x_{p_{ij}} / \sum_{i,j \in \{1,2,3\}} w_{p_{ij}}, \\ y_{c_{123}} = \sum_{i,j \in \{1,2,3\}} w_{p_{ij}} y_{p_{ij}} / \sum_{i,j \in \{1,2,3\}} w_{p_{ij}}, \end{cases} \tag{5.34}$$

where $w_{p_{i,j}} = \frac{1}{r_i + r_j}$, $(i, j \in \{1, 2, 3\})$ is the corresponding weighting factor that is inversely proportional to the distance between the intersection and the centres of three circles.

However, solely relying on a single combination of three circles is susceptible to inaccurate measurements collected by a certain Bluetooth gateway. Thus, measurements from more than three gateways are used to increase the system's redundancy, which is called "multiple weighted centroid localization". More concretely, we select the top K gateways whose measured RSSIs are strongest among all gateways, then select any three of them to make a group and estimate the device's position. Therefore, we have C_K^3 estimated positions, where the notation C_K^3 here denotes the number of combinations picking any 3 circles out of K circles. As illustrated in Fig. 5.57a, assume $K = 4$, there are $C_K^3 = C_4^3 = 4$ combinations of circles, i.e., (c_1, c_2, c_3), (c_1, c_2, c_4), (c_1, c_3, c_4) and (c_2, c_3, c_4). Accordingly, four centroids, i.e.,

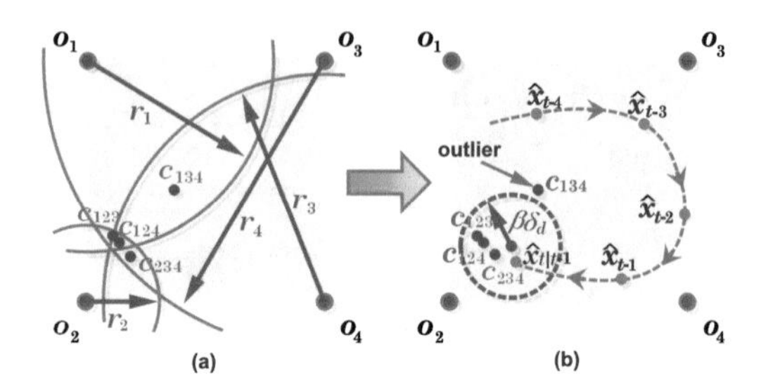

Fig. 5.57 Illustration of multiple weighted centroid approach. (**a**) Calculate multiple centroids. (**b**) Remove outliers

$c_{123} : (x_{c_{123}}, y_{c_{123}})$, $c_{124} : (x_{c_{124}}, y_{c_{124}})$, $c_{134} : (x_{c_{134}}, y_{c_{134}})$ and $c_{234} : (x_{c_{234}}, y_{c_{234}})$, can be obtained, then the averaged centroid $\overline{c} : (\overline{x}_c, \overline{y}_c)$ is computed as

$$\begin{cases} \overline{x}_c = (x_{c_{123}} + x_{c_{124}} + x_{c_{134}} + x_{c_{234}})/4, \\ \overline{y}_c = (y_{c_{123}} + y_{c_{124}} + y_{c_{134}} + y_{c_{234}})/4. \end{cases} \tag{5.35}$$

Compared to trusting a single estimation only, averaging multiple estimations is more resistant to high variance of measurements. Moreover in implementation, some useful tricks are used. One of them is that, the outliers in group of obtained centroids should be removed before further processing. For example, as illustrated in Fig. 5.57b, we first define a reference point $\overline{r} : (\overline{x}_r, \overline{y}_r)$ as follows:

$$\begin{cases} \overline{x}_r = (x_{c_{123}} + x_{c_{124}} + x_{c_{134}} + x_{c_{234}} + \widehat{x}_{t|t-1})/5, \\ \overline{y}_r = (y_{c_{123}} + y_{c_{124}} + y_{c_{134}} + y_{c_{234}} + \widehat{y}_{t|t-1})/5, \end{cases} \tag{5.36}$$

where $\widehat{p}_{t|t-1} : (\widehat{x}_{t|t-1}, \widehat{y}_{t|t-1})$ is the predicted state at time t by Kalman tracking, which will be elaborated in the next subsection. Then, the distance from $\overline{r} : (\overline{x}_r, \overline{y}_r)$ to separate centroid $c_{ijk} : (x_{c_{ijk}}, y_{c_{ijk}})$, i.e., $d_{\overline{r}-c_{ijk}}$ $(i, j, k \in \{1, 2, \ldots, K\})$, are calculated. If $d_{\overline{r}-c_{ijk}} > \beta\delta_d$ (δ_d is the standard derivation of these calculated distances and β is an empirical coefficient), the corresponding obtained centroid $c_{ijk} : (x_{c_{ijk}}, y_{c_{ijk}})$ is regarded as the outlier and will be excluded from the obtained centroid list. As demonstrated in Fig. 5.57b, since $d_{\overline{r}-c_{134}} > \beta\delta_d$, the centroid $c_{134} : (x_{c_{134}}, y_{c_{134}})$ obtained by the combination of circles (c_1, c_3, c_4) is eliminated, and $\overline{c} : (\overline{x}_c, \overline{y}_c)$ can be updated as follows:

$$\begin{cases} \overline{x}_c = (x_{c_{123}} + x_{c_{124}} + x_{c_{234}})/3, \\ \overline{y}_c = (y_{c_{123}} + y_{c_{124}} + y_{c_{234}})/3. \end{cases} \tag{5.37}$$

The averaged multiple weighted centroid $\overline{c} : (\overline{x}_c, \overline{y}_c)$ can be deemed as the estimated position at time t.

By adopting multiple weighted centroid localization, it is capable of handling hardware failure and inaccurate measurements of a single device. However, it is still unable to tackle unpredictable sudden fluctuations of measured RSSI during a short period of time. This is because the estimated position at any time t only relies on the current observation and is independent from the estimations prior to t.

In order to make the system immune to sudden changes of radio environment, we utilize Kalman filtering (KF) tracking to optimize the predicted trajectory, which not only takes current but also multiple prior observations into consideration. KF is a dynamic model that controls new inputs to the system at each time step, and estimates the system state by jointly considering multiple sequential measurements (i.e., RSSIs). KF tracking is a recursive process, in which two distinct steps: *prediction* and *correction*, are operated alternately. Since localization is two-dimensional estimation (i.e., the x-axis and the y-axis), let's take x-axis as an example to illustrate the KF tracking process.

In predication phase, it predicts the system's state $\widehat{x}_{t|t-1}$ conditioned on prior time $t-1$ and the state error covariance matrix $P_{t|t-1}$ according to Eqs. (5.38) and (5.39), respectively.

$$\widehat{x}_{t|t-1} = A_{t-1}\widehat{x}_{t-1} + \omega_t, \tag{5.38}$$

$$P_{t|t-1} = A_{t-1}P_{t-1}A_{t-1}^T + Q_{t-1}, \tag{5.39}$$

where A and Q correspond to the state transition matrix and the process noise matrix, ω ($\omega \sim \mathcal{N}(0, Q)$) is the process noise, and the superscript T denotes the transpose operation.

In correction phase, Kalman gain K_t is calculated in the first place, which is used to update the predicted state $\widehat{x}_{t|t-1}$ and the state error covariance P_t at current time t. Noted that the estimated x coordinate $\widehat{x}_t$ is obtained by adding an adjustment (correction) to $\widehat{x}_{t|t-1}$ which is based on previous observations. More specifically, in Eq. (5.42), $z_t - H_t\widehat{x}_{t|t-1}$ represents the difference between the new observed data z_t (i.e., $z_t = \overline{x}_c$) and the predicted measured results $H_t\widehat{x}_{t|t-1}$.

$$K_t = P_{t|t-1}H_t^T(H_tP_{t|t-1}H_t^T + R_t)^{-1}, \tag{5.40}$$

$$P_t = (I - K_tH_t)P_{t|t-1}, \tag{5.41}$$

$$\widehat{x}_t = \widehat{x}_{t|t-1} + K_t(z_t - H_t\widehat{x}_{t|t-1}), \tag{5.42}$$

where H, R and I refer to the observation matrix, the measurement noise covariance matrix, and the identity matrix, respectively.

It is worth mentioning that Q and R are two diagonal matrices, whose values represent how much the estimated results trust the predicted state and the new observations, respectively. In addition to basic KF, in online estimation, we introduce a self-adaptive mechanism that we continuously monitor the change of the variance of a few latest estimated positions (say 10 steps) prior to current estimation, and dynamically tune the coefficients of the diagonal matrices Q and R accordingly. In concrete details, we increase the coefficient of R and reduce that of Q if the consecutive observations are suffering higher fluctuation, and decrease R and increase Q if recent measurements becomes less fluctuated. Empirically, the adjustment range varies from 0.02 to 0.2 for Q and from 0.1 to 1.0 for R, respectively.

Moreover, in real implementation, we adopt the unscented Kalman filter (UKF), a more advanced version of Kalman filter, since most of the time, people or devices do not move at constant velocity while UKF can handle the nonlinear system better.

5.2.1.2 Experiments

Experiment Setup In experiment, we deploy our system on a remote cloud server, and Bluetooth gateways are mounted on the ceiling of lab room. During the test, tester is walking back and forth in the testing area with the transmitting tag fixed on his helmet. In deployment, one of the useful tips is that four gateways should be placed at four corners of the lab room in a square shape and it is also recommend to place another gateway in the centre of the room.

We use the root mean square error (RMSE) to evaluate the performance of our localization system, i.e.,

$$\text{RMSE} = \sqrt{\frac{1}{T}\sum\nolimits_{t=1}^{T}\left[(\widehat{x}_t - \widetilde{x}_t)^2 + (\widehat{y}_t - \widetilde{y}_t)^2\right]}, \tag{5.43}$$

where $t \in \{1, 2, \ldots, T\}$ is the time step, and $(\widetilde{x}_t, \widetilde{y}_t)$ are the ground truth at time t.

Comparison of Different Distance Estimator Figure 5.58 compares prediction accuracy of our proposed LSTM-based distance estimator with the one using conventional propagation model in terms of RMSE. In evaluation, we first collect training labels for every testing point that is n meters ($n \in \{1, 2, \ldots, 8\}$) away from an anchor point. Then, the collected data are used to train the LSTM-based distance estimator using supervised learning method. Results show that the average localization error starts from 0.24 and 0.15 m for propagation model-based estimator and LSTM-based estimator, respectively, when the distance between tester and gateway is exactly 1, to 0.96 m and 0.62 m as the distance goes to 8 m, respectively,. This is because compared to theoretical propagation model, our proposed LSTM-based estimator is fully based on deep learning models, which is much more powerful in modelling the complicated non-linear attenuation of radio signal.

Comparison of Different Localization Algorithms Table 5.4 presents the performance of our proposed localization algorithm (i.e., LSTM + Tri (Multi-Weighted-Centroid) + KF) as well as other the state-of-the-art benchmarks. Least square is one of the popular method used to estimate the optimal solution when solving geometry equations in pure RSSI-based trilateral localization. As we can see, its performance is at the bottom of the ranking no matter it cooperates with either propagation model-based estimator or LSTM-based estimator. This is because least square is less likely to resistent to inaccurate measurements, even though it can reach good result in ideal

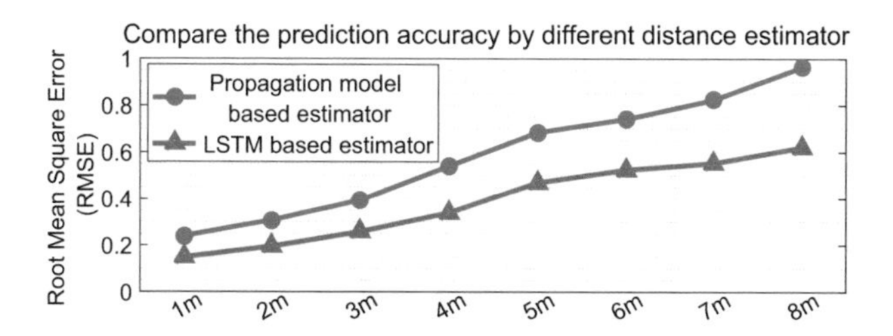

Fig. 5.58 Comparison of different distance estimators

Table 5.4 Comparison of the localization error across different localization algorithms in terms of RMSE

Localization algorithm	Localization err.
Prop + Tri (least square)	1.9319 m
LSTM + Tri (least square)	1.8809 m
Prop + Tri (multi-weighted-centroid)	1.2354 m
LSTM + Tri (multi-weighted-centroid)	1.1598 m
LSTM + Tri (multi-weighted-centroid) + KF	0.8650 m

radio environment. Weighted centroid trilateration estimates the relative position of the device via geometry calculation of three circles. Multiple weighted centroid localization (Multi-Weighted-Centroid) averages multiple centroids obtained by different combinations of any three circles, which performs much better over Least Square. Our proposed LSTM + Tri (Multi-Weighted-Centroid) + KF can achieve around 0.86 m localization accuracy that surpasses all other baselines by a large margin.

Performance in Large Scale Warehouse Scenario In addition to lab environment, we also deploy and test our indoor localization system in Alibaba's Tianmao warehouse. Figure 5.59 demonstrates how we deploy Bluetooth localization devices that user tag is affixed to tester's helmet, and Bluetooth gateways are mounted on top of the storage rack receiving Bluetooth signals sent by user tags. During the test, tester randomly walks along aisles between two racks in the warehouse, and the estimated real-time trajectory are computed and visualized on Web interface. According to statistics on the back-end, our system reaches around 1.5 m localization accuracy in RMSE. Compared to ideal LAB environment, the performance in logistics warehouse scenario deteriorates to some extent because wireless signals are reflected, diffracted or even absorbed by dense metal bars of storage racks and also blocked by huge amounts of stacked goods.

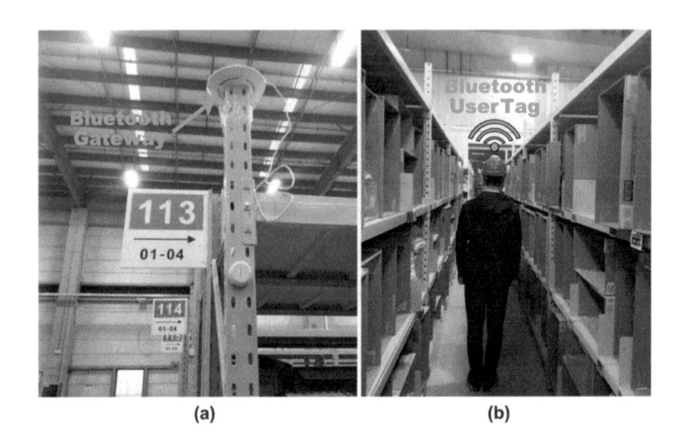

Fig. 5.59 Demonstration of the deployment of Bluetooth devices in Alibaba's Tianmao warehouse. (**a**) Bluetooth gateway mounted on top of storage rack. (**b**) Bluetooth user tag affixed to tester's helmet

5.2.2 RFID-Based Object Indoor Localization

In the future smart warehousing scenarios, robots may completely replace human beings in terms of automatic object fetching and delivery. Indoor localization is one of the most important techniques for realizing this vision. In fact, a batch of techniques, such as GPS, wireless sensor, bluetooth, Wi-Fi and computer vision, have been proposed. However, none of these techniques is suitable for large-scale warehousing scenarios due to the following reasons: (1) The GPS technique works well for the outdoor localization and navigation, but fails in the indoor scenarios; (2) The techniques based on wireless sensors and bluetooth beacons cannot provide a long-term localization service due to the limited volume of batteries; (3) The WiFi-based tracking techniques exploit signal reflection to locate objects but fail to distinguish similar objects; (4) The computer vision-based approaches require line-of-sight between the target objects and camera.

Compared with the above techniques, RFID naturally has various advantages including low cost, easy deployment, battery-free, individual identification and no requirement on line-of-sight. Hence, in this case study, we investigate the RFID technique for object localization in large-scale warehousing scenarios.

5.2.2.1 Understanding and Preprocessing Phase

We suppose that the RFID reader has received n readings from the target tag with id. Thus, we have n phase points in the raw phase profile: $\mathcal{P}(id, t_1), \mathcal{P}(id, t_2), ..., \mathcal{P}(id, t_n)$ while the timestamps $t_1, t_2, ..., t_n$ are in an ascending order, i.e., for any $1 \leq i < j \leq n$, we have $t_i < t_j$. We use $dis(id, t_i)$

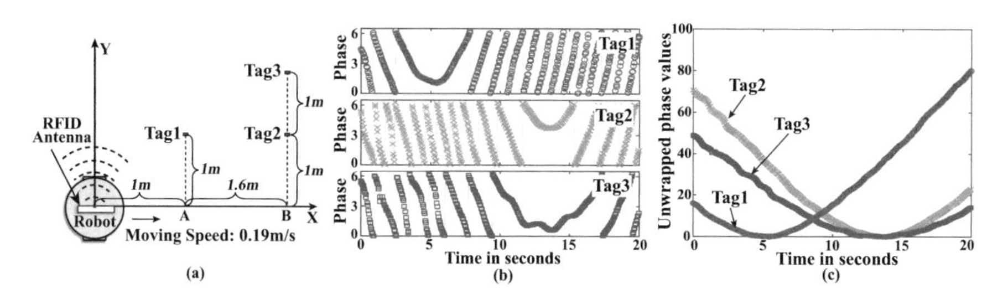

Fig. 5.60 Processing phase profile: (**a**) illustrating the experiment deployment, (**b**) raw phase profiles, (**c**) unwrapped phase profiles

to denote the distance between the reader antenna and the tag id at time t_i. The signal traverses a total distance of $2 \times dis(id, t_i)$ back and forth in backscatter communication. Besides phase rotation over distance, the reader's transmitter circuits, the tag's reflection characteristic, and the reader's receiver circuits will also introduce some additional phase rotations, denoted as φ_T, φ_{TAG} and φ_R respectively. The phase value $\mathcal{P}(id, t_i)$ returned by the RFID reader can be expressed as follows:

$$\mathcal{P}(id, t_i) = \left[\frac{2 \times dis(id, t_i)}{\lambda} \times 2\pi + \Theta\right] \mod 2\pi, \tag{5.44}$$

where λ is the wavelength of the RFID signal, and the constant Θ, called hardware diversity, equals $\varphi_T + \varphi_{TAG} + \varphi_R$.

Next, we conduct a set of experiments to better understand the phase profile. As illustrated in Fig. 5.60a, we deploy three slim RFID tags vertically in the system. The moving speed v of the robot is set to 0.19 m/s. The RFID reader keeps interrogating tags during the moving process, and the collected raw phase profiles of these three tags are plotted in Fig. 5.60b, respectively.

The raw phase profile of each tag involves the following two types of phase noises. (i) *Random error*: The authors of [39] conducted an empirical study over 100 tags with environment temperature from 0 to 40 °C, and pointed out that phase measurement results inevitably contain random errors, following a typical Gaussian distribution with a standard deviation of 0.1 radians. (ii) *Periodic jump*: according to Eq. (5.44), the tag phase is a periodic function that repeats if the distance between the reader antenna and tag changes by $\lambda/2$. We first investigate how to remove periodic jumps from the phase profile, and will take random errors into consideration when quantifying the deviation of localization results.

As shown in Fig. 5.60b, the raw tag phase profile involves periodic phase jumps due to the `mod` operation in Eq. (5.44). These phase jumps are either from a phase value around 0 to a follow-up phase value around 2π or from a phase value around 2π to a follow-up phase value around 0. We can use a method similar with the `unwrap` command in Matlab to remove the phase jumps in the phase profile $\mathcal{P}(id, t_1), \mathcal{P}(id, t_2), \cdots, \mathcal{P}(id, t_n)$ by pulsing or minusing multiples of 2π

when the absolute phase jumps between consecutive phase values are greater than or equal to the default jump tolerance. Using such a method, we can remove the impact of mod operation and obtain a new sequence of unwrapped phase values: $\mathcal{P}'(id, t_1), \mathcal{P}'(id, t_2), \cdots, \mathcal{P}'(id, t_n)$, which looks like a shape of V. Specifically, an arbitrary phase point $\mathcal{P}'(id, t_i)$ in the unwrapped phase profile can be expressed as follows:

$$\mathcal{P}'(id, t_i) = \frac{2 \times dis(id, t_i)}{\lambda} \times 2\pi + \Theta + 2k\pi, \tag{5.45}$$

where k is a constant integer within $\{0, \pm 1, \pm 2, \cdots\}$. Next section will use the unwrapped phase profile to calculate tag location.

5.2.2.2 Detailed Design for 2D Localization

For a target tag on the X-Y plane, we still suppose the reader has received its n replies. Thus, we have n unwrapped phase points after unwrapping operations: $\mathcal{P}'(id, t_1), \mathcal{P}'(id, t_2), \cdots, \mathcal{P}'(id, t_n)$. The proposed system equally partitions the n unwrapped phase points into three segments: $[\mathcal{P}'(id, t_1), \cdots, \mathcal{P}'(id, t_w)]$, $[\mathcal{P}'(id, t_{w+1}), \cdots, \mathcal{P}'(id, t_{2w})]$, $[\mathcal{P}'(id, t_{2w+1}), \cdots, \mathcal{P}'(id, t_{3w})]$, where $w = \lfloor \frac{n}{3} \rfloor$. Then, we take the i-th phase value $\mathcal{P}'(id, t_i)$ from the first segment, the i-th phase value $\mathcal{P}'(id, t_{w+i})$ from the second segment, and the i-th phase value $\mathcal{P}'(id, t_{2w+i})$ from the third segment, where $i \in [1, w]$. Next, we will describe how to use these three picked phase values to calculate the location of target tag. Since there are w such phase triads, the system can calculate w candidate tag locations. To distinguish these candidate tag locations from each other, we use (x_i, y_i) to denote the candidate tag location calculated from the phase triad: $\mathcal{P}'(id, t_i)$, $\mathcal{P}'(id, t_{w+i})$, and $\mathcal{P}'(id, t_{2w+i})$.

As exemplified in Fig. 5.61, we assume that the reader antenna arrives at the locations I, J, K at the time points of t_i, t_{w+i}, t_{2w+i}, respectively. According to Eq. (5.45), we can calculate the difference between adjacent phase points $\mathcal{P}'(id, t_i)$ and $\mathcal{P}'(id, t_{w+i})$, and the difference between adjacent phase points $\mathcal{P}'(id, t_{w+i})$ and $\mathcal{P}'(id, t_{2w+i})$ as follows:

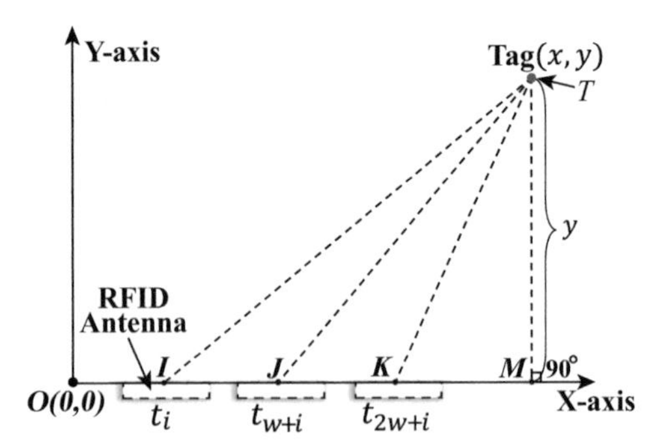

Fig. 5.61 Exemplifying the principle for 2D localization

$$
\begin{aligned}
\mathcal{P}'(id, t_i) - \mathcal{P}'(id, t_{w+i}) &= \frac{4\pi \times \left(|\overrightarrow{IT}| - |\overrightarrow{JT}|\right)}{\lambda}. \\
\mathcal{P}'(id, t_{w+i}) - \mathcal{P}'(id, t_{2w+i}) &= \frac{4\pi \times \left(|\overrightarrow{JT}| - |\overrightarrow{KT}|\right)}{\lambda}.
\end{aligned}
\tag{5.46}
$$

According to the geometric relationships shown in Fig. 5.61, we also have the following equations:

$$
\begin{cases}
|\overrightarrow{IT}| = \sqrt{|\overrightarrow{IM}|^2 + |\overrightarrow{MT}|^2}, \\
\overrightarrow{IM} = \overrightarrow{IK} + \overrightarrow{KM}, \\
|\overrightarrow{JT}| = \sqrt{|\overrightarrow{JM}|^2 + |\overrightarrow{MT}|^2}, \\
\overrightarrow{JM} = \overrightarrow{JK} + \overrightarrow{KM}, \\
|\overrightarrow{KT}| = \sqrt{|\overrightarrow{KM}|^2 + |\overrightarrow{MT}|^2}, \\
\overrightarrow{KM} = \overrightarrow{OM} - \overrightarrow{OK}, \\
\overrightarrow{IK} = [v(t_{2w+i} - t_i), 0], \\
\overrightarrow{JK} = [v(t_{2w+i} - t_{w+i}), 0], \\
\overrightarrow{OK} = (vt_{2w+i}, 0), \\
\overrightarrow{MT} = (0, y), \\
\overrightarrow{OM} = (x, 0).
\end{cases}
$$

By substituting the above equations into Eq. (5.46), we obtain an equation set that contains two unknown variables x and y. Then, we solve the equation set to get the candidate tag location (x_i, y_i).

$$
\begin{aligned}
x_i &\leftarrow x = vt_{2w+i} + \frac{(\frac{\lambda \Delta\varphi_2}{4\pi})^2 - v^2 \Delta T_2^2 + \frac{\lambda \Delta\varphi_2 \mathcal{S}}{2\pi}}{2v\Delta T_2}, \\
y_i &\leftarrow y = \sqrt{\mathcal{S}^2 - (x_i - vt_{2w+i})^2},
\end{aligned}
\tag{5.47}
$$

where the values of $\Delta\varphi_2$, ΔT_2, and $\mathcal{S}$ are as follows:

$$\begin{cases} \mathcal{S}=\left\{v^2(\Delta T_1+\Delta T_2)^2+\left(\frac{\Delta T_1}{\Delta T_2}+1\right)\left[\left(\frac{\lambda\Delta\varphi_2}{4\pi}\right)^2-v^2\Delta T_2^2\right],\right. \\ \qquad \left.-\left[\frac{\lambda(\Delta\varphi_1+\Delta\varphi_2)}{4\pi}\right]^2\right\}\Big/\left(\frac{\lambda\Delta\varphi_1}{2\pi}-\frac{\lambda\Delta\varphi_2\Delta T_1}{2\pi\Delta T_2}\right), \\ \Delta\varphi_1=\mathcal{P}'(id,t_i)-\mathcal{P}'(id,t_{w+i}), \\ \Delta\varphi_2=\mathcal{P}'(id,t_{w+i})-\mathcal{P}'(id,t_{2w+i}), \\ \Delta T_1=t_{w+i}-t_i, \\ \Delta T_2=t_{2w+i}-t_{w+i}. \\ \vdots \end{cases}$$

Due to the noise of random errors, the unwrapped phase value has a variance of $Var[\mathcal{P}'(id,t_i)]=0.01$. Then, the variances of $\Delta\varphi_1$ and $\Delta\varphi_2$ can be calculated as follows:

$$Var(\Delta\varphi_1)=Var[\mathcal{P}'(id,t_i)]+Var[\mathcal{P}'(id,t_{w+i})]=0.02,$$
$$Var(\Delta\varphi_2)=Var[\mathcal{P}'(id,t_{w+i})]+Var[\mathcal{P}'(id,t_{2w+i})]=0.02.$$

The probabilistic deviation inherent in $\Delta\varphi_1$ and $\Delta\varphi_2$ also results in that the candidate tag location (x_i, y_i) derived from Eq. (5.47) is also inaccurate. To quantify the localization deviation, we calculate the variance of x_i and y_i in the following. We observe from Eq. (5.47) that both x_i and y_i are functions of $\Delta\varphi_1$ and $\Delta\varphi_2$. Hence, we denote x_i as $\eta_x(\Delta\varphi_1,\Delta\varphi_2)$ and y_i as $\eta_y(\Delta\varphi_1,\Delta\varphi_2)$, respectively. We present the Taylor's series expansion of x_i and y_i around (h_1,h_2), respectively. Here, $h_1=E(\Delta\varphi_1)$ and $h_2=E(\Delta\varphi_2)$.

$$x_i=\eta_x(h_1,h_2)+\frac{\partial\eta_x}{\partial\Delta\varphi_1}(\Delta\varphi_1-h_1)+\frac{\partial\eta_x}{\partial\Delta\varphi_2}(\Delta\varphi_2-h_2),$$
$$y_i=\eta_y(h_1,h_2)+\frac{\partial\eta_y}{\partial\Delta\varphi_1}(\Delta\varphi_1-h_1)+\frac{\partial\eta_y}{\partial\Delta\varphi_2}(\Delta\varphi_2-h_2).$$

We have the following equation by taking expectation of both sides of the above two equations, respectively.

$$\begin{aligned} E(x_i)&=\eta_x(h_1,h_2), \\ E(y_i)&=\eta_y(h_1,h_2). \end{aligned} \tag{5.48}$$

With Eq. (5.48), we can calculate the variance of x_i and y_i as follows:

$$
\begin{aligned}
Var(x_i) &= E[x_i - E(x_i)]^2 \\
&= (\frac{\partial \eta_x}{\partial \Delta\varphi_1})^2 Var(\Delta\varphi_1) + (\frac{\partial \eta_x}{\partial \Delta\varphi_2})^2 Var(\Delta\varphi_2), \\
Var(y_i) &= E[y_i - E(y_i)]^2 \\
&= (\frac{\partial \eta_y}{\partial \Delta\varphi_1})^2 Var(\Delta\varphi_1) + (\frac{\partial \eta_y}{\partial \Delta\varphi_2})^2 Var(\Delta\varphi_2).
\end{aligned} \tag{5.49}
$$

As required in Eq. (5.49), we need to calculate the expressions of $\frac{\partial \eta_x}{\partial \Delta\varphi_1}$, $\frac{\partial \eta_x}{\partial \Delta\varphi_2}$, $\frac{\partial \eta_y}{\partial \Delta\varphi_1}$, and $\frac{\partial \eta_y}{\partial \Delta\varphi_2}$, respectively.

Due to the complexity of the expressions, we use some symbols to denote the terms that repetitively appear in equation. Specifically, $\mathcal{A} = v^2(\Delta T_1 + \Delta T_2)^2$, $\mathcal{B} = \frac{\Delta T_1}{\Delta T_2} + 1$, $\mathcal{C} = v\Delta T_2$, $\mathcal{D} = \frac{\lambda\Delta\varphi_1}{4\pi}$, $\mathcal{E} = \frac{\lambda\Delta\varphi_2}{4\pi}$, $\mathcal{F} = \frac{\lambda(\Delta\varphi_1+\Delta\varphi_2)}{4\pi}$. Then, the expressions of $\frac{\partial \eta_x}{\partial \Delta\varphi_1}$ and $\frac{\partial \eta_x}{\partial \Delta\varphi_2}$ are given as follows:

$$
\begin{aligned}
\frac{\partial \eta_x}{\partial \Delta\varphi_1} &= \frac{-2\mathcal{E}\mathcal{F}(\Delta\varphi_1-(\mathcal{B}-1)\Delta\varphi_2)-\Delta\varphi_2[\mathcal{A}+\mathcal{B}(\mathcal{E}^2-\mathcal{C}^2)-\mathcal{F}]}{2\mathcal{C}[\Delta\varphi_1-(\mathcal{B}-1)\Delta\varphi_2]^2}, \\
\frac{\partial \eta_x}{\partial \Delta\varphi_2} &= \frac{\lambda^2\Delta\varphi_2}{\mathcal{C}(4\pi)^2} + \frac{\mathcal{G}-\mathcal{H}}{2\mathcal{C}[\Delta\varphi_1-(B-1)\Delta\varphi_2]^2},
\end{aligned} \tag{5.50}
$$

where the expressions of $\mathcal{G}$ and $\mathcal{H}$ are given below.

$$
\begin{cases}
\mathcal{G} = [\mathcal{A}+\mathcal{B}(\mathcal{E}^2-\mathcal{C}^2)-\mathcal{F}^2+\Delta\varphi_2(\frac{2\mathcal{B}\mathcal{E}^2}{\Delta\varphi_2}-\frac{\mathcal{F}\lambda}{2\pi})][\Delta\varphi_1-(\mathcal{B}-1)\Delta\varphi_2], \\
\mathcal{H} = \Delta\varphi_2[\mathcal{A}+\mathcal{B}(\mathcal{E}^2-\mathcal{C}^2)-\mathcal{F}^2](1-\mathcal{B}).
\end{cases}
$$

And the expressions of $\frac{\partial \eta_y}{\partial \Delta\varphi_1}$ and $\frac{\partial \eta_y}{\partial \Delta\varphi_2}$ are given as follows:

$$
\begin{aligned}
\frac{\partial \eta_y}{\partial \Delta\varphi_1} &= \frac{2\mathcal{S}\frac{\partial \mathcal{S}}{\partial \Delta\varphi_1} - 2(x_i - vt_{2w+i})\frac{\partial \eta_x}{\partial \Delta\varphi_1}}{2\sqrt{\mathcal{S}^2-(x_i - vt_{2w+i})^2}}, \\
\frac{\partial \eta_y}{\partial \Delta\varphi_2} &= \frac{2\mathcal{S}\frac{\partial \mathcal{S}}{\partial \Delta\varphi_2} - 2(x_i - vt_{2w+i})\frac{\partial \eta_x}{\partial \Delta\varphi_2}}{2\sqrt{\mathcal{S}^2-(x_i - vt_{2w+i})^2}},
\end{aligned} \tag{5.51}
$$

in which $\frac{\partial \mathcal{S}}{\partial \Delta\varphi_1}$ and $\frac{\partial \mathcal{S}}{\partial \Delta\varphi_2}$ are calculated as follows:

$$
\begin{cases}
\frac{\partial \mathcal{S}}{\partial \Delta\varphi_1} = \frac{-2\mathcal{F}\lambda[\mathcal{D}-(\mathcal{B}-1)\mathcal{E}]-\lambda\{[\mathcal{A}+\mathcal{B}(\mathcal{E}^2-\mathcal{C}^2)]-\mathcal{F}^2\}}{8\pi[\mathcal{D}-(\mathcal{B}-1)\mathcal{E}]^2}, \\
\frac{\partial \mathcal{S}}{\partial \Delta\varphi_2} = \frac{2\lambda(\mathcal{B}\mathcal{E}-\mathcal{F})[\mathcal{D}-(\mathcal{B}-1)\mathcal{E}]-\lambda[\mathcal{A}+\mathcal{B}(\mathcal{E}^2-\mathcal{C}^2)-\mathcal{F}^2](1-\mathcal{B})}{8\pi[\mathcal{D}-(\mathcal{B}-1)\mathcal{E}]^2}.
\end{cases}
$$

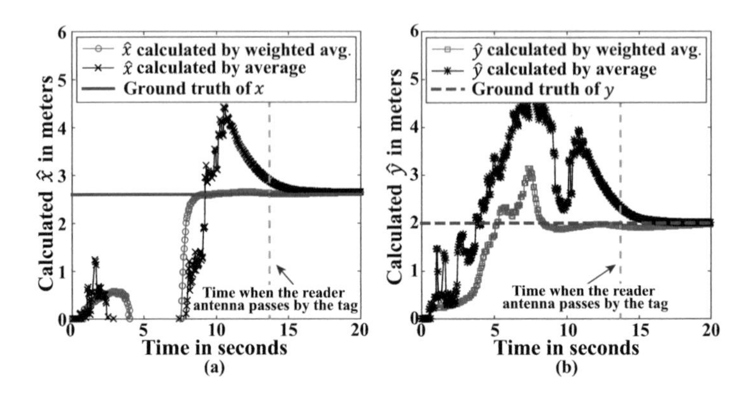

Fig. 5.62 Comparison between weighted average results and direct average results: (**a**) calculated $\hat{x}$ vs. time; (**b**) calculated $\hat{y}$ vs. time

So far, we have calculated the candidate location of the target tag, i.e., (x_i, y_i) in Eq. (5.47), as well as their variances in Eq. (5.49). Recall that the proposed system can calculate w candidate tag locations: (x_1, y_1), (x_2, y_2), ..., (x_w, y_w). A straightforward way is to directly use their average as the final localization result. It is simple but far from optimal, because candidate tag locations have different variances. Intuitively, if all three picked phase points lie in the very left part of the unwrapped phase profile (nearly in a straight line), the calculated candidate tag location may be not very accurate. Hence, instead of directly using the average of candidate tag locations, we use their weighted average as the final localization result. A candidate tag location with a smaller variance should be assigned with a larger weight, and vice versa. Hence, we use $\frac{1}{Var(x_i)}$ as the weight of x_i, and use $\frac{1}{Var(y_i)}$ as the weight of y_i. Then, we calculate the final tag location by $\hat{x} = \sum_{i=1}^{w} \frac{x_i}{Var(x_i)\aleph_x}$ and $\hat{y} = \sum_{i=1}^{w} \frac{y_i}{Var(y_i)\aleph_y}$, where $\aleph_x = \sum_{i=1}^{w} \frac{1}{Var(x_i)}$ and $\aleph_y = \sum_{i=1}^{w} \frac{1}{Var(y_i)}$.

In Fig. 5.62a, b, we plot the values $\hat{x}$ and $\hat{y}$ that are calculated by the direct average results and the weighted average results, respectively. We make two main observations from the experimental results. First, the weighted average method is faster to converge to the ground truth than the simple average method. Second, the values of $\hat{x}$ and $\hat{y}$ have been already very close to the ground truth at the 7-th second, which is 6 s earlier than the time when the reader antenna passes by the target tag. Since the robot speed is set to 20 cm/s, it means that can achieve relatively accurate localization result about 1.2 meters before reader antenna passes by the target tag. In other words, the system is able to locate the tagged objects in corner where robot cannot pass by.

5.2.2.3 Extending to 3D Localization

The proposed system can be easily extended to enable 3D localization by simultaneously using two reader antennas $\mathcal{R}_1$ and $\mathcal{R}_2$. As illustrated in Fig. 5.63, we suppose the distance between two reader antennas is h meters. Since the below antenna starts at the point $O(0, 0, 0)$, the above antenna will start at the point $O'(0, 0, h)$. The trajectories of $\mathcal{R}_1$ and $\mathcal{R}_2$ are parallel to each other, and also with a distance of h. As aforementioned, we can leverage the antenna port information in each tag reading to distinguish which antenna the current tag reading is received from. Thus, we can have two phase profiles of the target tag corresponding to these two reader antennas, respectively.

As illustrated in Fig. 5.63, we have two planes in the 3D space: TOM_1 and $TO'M_2$. On the plane TOM_1, the 2D localization approach described in the above is applied on the phase profile corresponding to antenna $\mathcal{R}_1$ and we can calculate a tag location $(\hat{x}_1, \hat{y}_1)$. We draw a line from tag location $T(x, y, z)$ perpendicularly to the trajectory of $\mathcal{R}_1$, with foot M_1. On the plane of $T\mathcal{O}M_1$, it is easy to know that $|\overrightarrow{\mathcal{O}M_1}| = \hat{x}_1$ and $|\overrightarrow{TM_1}| = \hat{y}_1$. Similarly, applying the 2D localization approach on the phase profile from $\mathcal{R}_2$, the system can also calculate a tag location $(\hat{x}_2, \hat{y}_2)$, which satisfy that $|\overrightarrow{\mathcal{O}'M_2}| = \hat{x}_2$ and $|\overrightarrow{TM_2}| = \hat{y}_2$. It is easy to know that the coordinates of M_1 and M_2 are $(\hat{x}_1, 0, 0)$ and $(\hat{x}_2, 0, h)$, respectively. In the ideal case, we should have $x = \hat{x}_1 = \hat{x}_2$. Due to the deviation in localization results, $\hat{x}_1$ may not exactly equal $\hat{x}_2$. Then, we calculate the coordinate value $x = \frac{\hat{x}_1+\hat{x}_2}{2}$.

Next, we investigate how to calculate the coordinate values y and z of the target tag. Three types of geometric relationships in the triangle ΔTM_1M_2 are illustrated in Fig. 5.64, which correspond to $z \in (0, h]$, $z \leq 0$, $z > h$, respectively. No matter which geometric relationship actually applies, we always have the following equation set:

$$\begin{cases} |\overrightarrow{TM_1}| = \sqrt{y^2 + z^2} = \hat{y}_1, \\ |\overrightarrow{TM_2}| = \sqrt{y^2 + (z - h)^2} = \hat{y}_2. \end{cases}$$

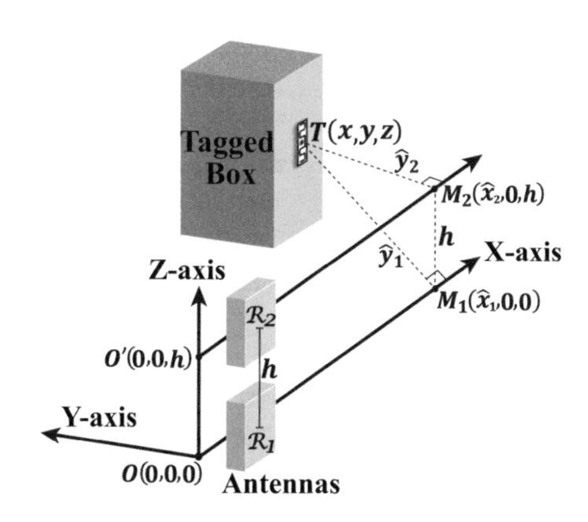

Fig. 5.63 3D localization

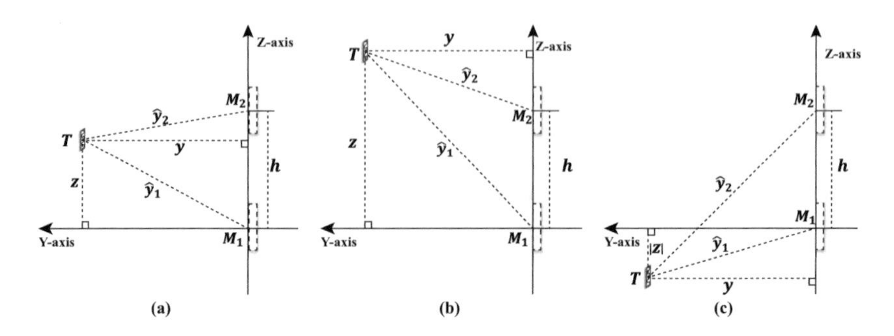

Fig. 5.64 Geometric relationship between variables y, z, $\hat{y}_1$, $\hat{y}_2$, and h. (**a**) $0 < z \leq h$. (**b**) $z > h$. (**c**) $z \leq 0$

By solving the above equation set, we can calculate the coordinates of the target tag in 3D space as follows:

$$\begin{cases} \hat{x} = \frac{\hat{x}_1+\hat{x}_2}{2}, \\ \hat{y} = \sqrt{(\hat{y}_1)^2 - [\frac{(\hat{y}_1)^2-(\hat{y}_2)^2+h^2}{2h}]^2}, \\ \hat{z} = \frac{(\hat{y}_1)^2-(\hat{y}_2)^2+h^2}{2h}. \end{cases}$$

So far, the system has been extended to successfully enable the 3D localization functionality.

5.2.2.4 Evaluation

Hardware Components The proposed system consists of the following hardware components: a Thinkpad Carbon X1 desktop, an EAI Dashgo D1 smart robot, an Impinj R420 reader, two Laird S9028PCL reader antennas, and several impinj e41c tags. RFID reader works at the UHF band 902∼928 MHz. To eliminate the channel hopping issue, we fix the working frequency at 920.625 MHz. We configure the reader transmission power to 32.5 dBm and use the circular polarization antennas in this paper. The circularly polarized Laird S9028PCL antenna is of the gain 8.5 dBic and also operates within 902–928 MHz. The RFID reader reports the low level RFID data to the desktop via a WiFi embedded in the robot. Following the experiment settings of the state-of-the-art RF-Scanner localization system, we also assume that there is a line-of-sight between reader antenna and tag. In this paper, we use the ALIEN 9640 tags, which are in the shape of strip. Moreover, the tags are placed vertically as default, i.e., the tag orientation is along the Z-axis.

Software Configuration The robot is controlled by an arduino board with a bluetooth communication module. Utilizing the Bluetooth channel, we have an application on the smart phone to control the robot movement with given direction

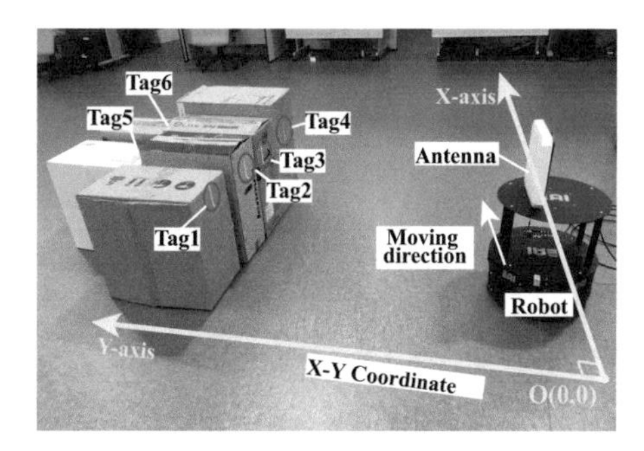

Fig. 5.65 Deployment of the MRL system in the 2D plane

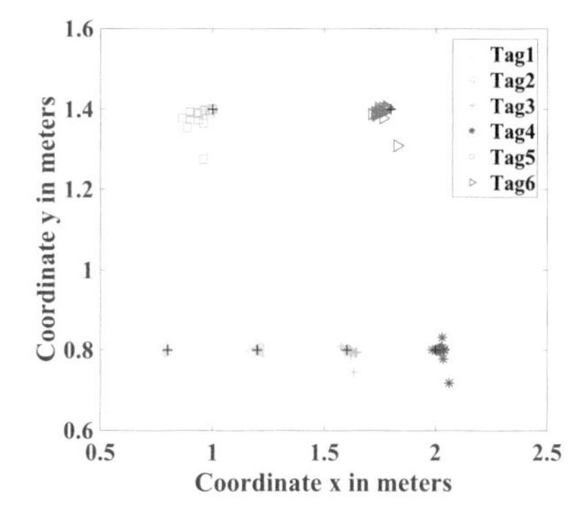

Fig. 5.66 Localization results of multiple RFID tags in 2D plane

and speed. On the server side, we first adopt the LLRP protocol, which is implemented in Java, to configure the reader to read the low level RFID data (e.g., tag ID, phase, timestamp) from tags. The collected RFID data are timely stored in a local file on the server. At the same time, MRL reads the data from this local file to calculate the tag location.

Performance in 2D Localization First, we investigate the localization accuracy of the MRL system in the 2D plane. As illustrated in Fig. 5.65, we attach 6 tags to 6 cartons. The tagged cartons are placed more than 0.8 m away from the trajectory of the moving robot. The actual tag locations are marked by "+" in Fig. 5.66. The reader antenna and these 6 tags are on the same plane. The MRL system passes by these tagged cartons with a speed $v = 0.1$ m/s and then reports their locations. The same experiment is repeated for several times, and the localization results are also plotted in Fig. 5.66.

We observe that the calculated locations for each tag are very close to the ground truth. For a tag with location (x, y), if the calculated location is $(\hat{x}, \hat{y})$, we refer to $|x - \hat{x}|$ as the localization error in $\hat{x}$, and similarly refer to $|y - \hat{y}|$ as the localization error in $\hat{y}$. For clearly evaluating the localization accuracy of MRL, we also plot the CDF curves of localization errors in $\hat{x}$ and $\hat{y}$ in Fig. 5.67. We observe from Fig. 5.67 that, the localization errors along X-axis are generally less than that along Y-axis. The results in Fig. 5.67a–d reveal that localization errors of the MRL system are less

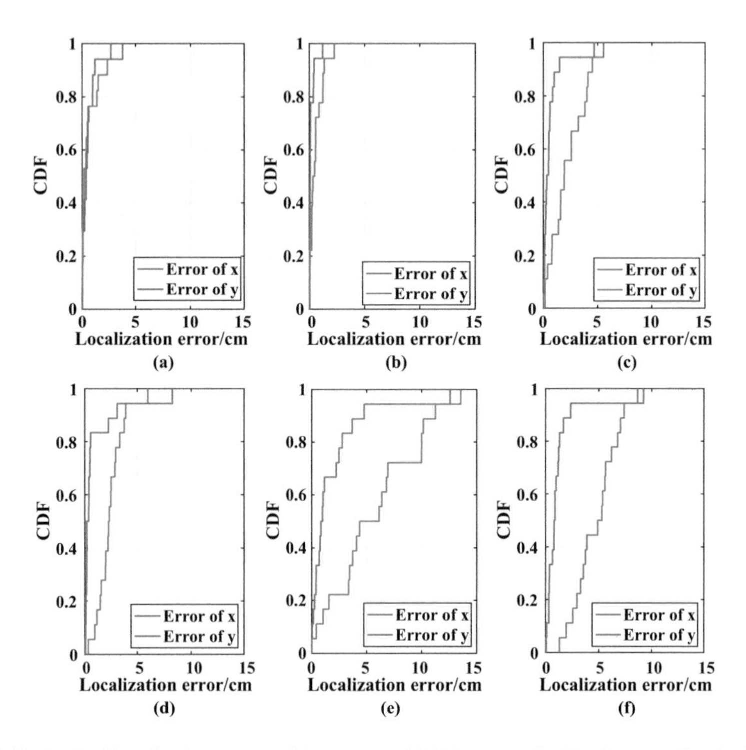

Fig. 5.67 CDF of localization errors of the proposed MRL system in 2D plane. (**a**) Tag 1. (**b**) Tag 2. (**c**) Tag 3. (**d**) Tag 4. (**e**) Tag 5. (**f**) Tag 6

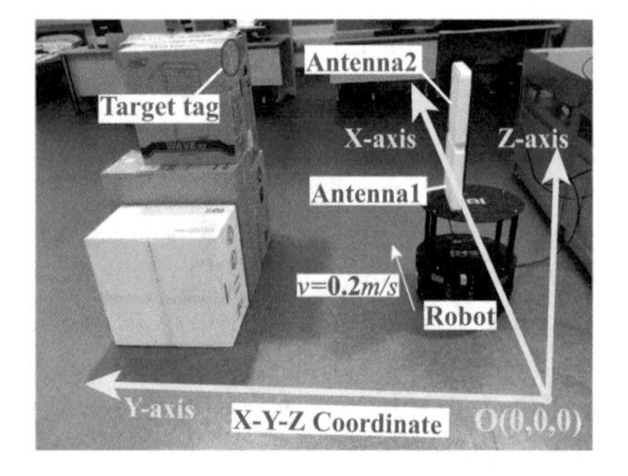

Fig. 5.68 Deployment of the MRL system in the 3D space

than 5 cm with a probability larger than 90%. However, localization errors of Tag 5 and Tag 6 are a bit larger than that of the other tags, because signals of these two tags are affected by the cartons in the line-of-sight path to the reader antenna.

Performance in 3D Localization As illustrated in Fig. 5.68, we fix two antennas with a distance $h = 0.7$ m on the robot. The moving trajectory of antenna $\mathcal{R}_1$ is treated as the positive direction of the X-axis. In such a coordinate system, we place a target tag at different positions. The ground truth of x varies from 1 to 1.5 m; the ground truth of y varies from 0.6 to 0.8 m and 1 m; the ground truth of z varies from

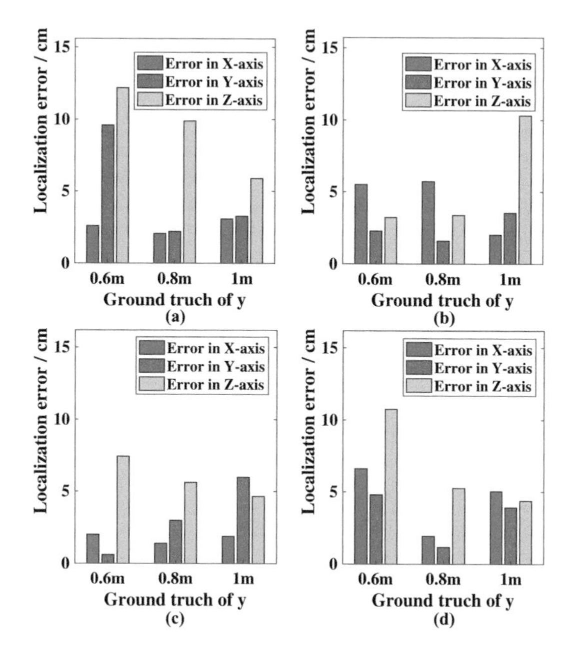

Fig. 5.69 Investigating localization accuracy of MRL vs. different tag positions. (**a**) tag is placed at (1 m, y, 0.35 m). (**b**) tag is placed at (1.5 m, y, 0.35 m). (**c**) tag is placed at (1 m, y, 0.7 m). (**d**) tag is placed at (1.5 m, y, 0.7 m). *y* varies from 0.6 to 0.8 m and 1 m

0.35 to 0.7 m. The MRL system moves with a speed of 10 cm/s and pinpoints the tag location.

The experimental results shown in Fig. 5.69 reveal the 3D localization accuracy of the proposed MRL system: most localization errors along the X- and Y-axes are less than 7 cm and most localization errors along the Z-axis are less than 12 cm. Such a localization accuracy can satisfy the requirements of most application scenarios.

5.3 Liquid Sensing

Liquid counterfeiting and adulteration have been jeopardizing human health for many decades. Fake liquids pose huge health risks and result in a large number of poisoning cases every year. Common liquid fraud involves adulterating the expensive authentic liquid with cheaper and even harmful liquids or counterfeiting the authentic liquid with a similar flavor but totally different components. Adulterated and counterfeiting liquids are difficult to detect for consumers, since they are camouflaged with the same appearance as the authentic one while only with the fake liquids inside. In recent years, governments, industries, and academia have taken great efforts to fight against liquid adulteration and counterfeiting. However, people are still suffering from high risks owing to the lack of efficient and ubiquitous liquid detection approaches. This drives researchers to keep investigating better methods to detect liquid fraud.

Existing solutions for liquid fraud detection mainly rely on chemical and chromatographic techniques [13]. These techniques enable precise liquid component identification and estimation, especially for detecting the target contaminants in the liquid. However, chemical tools and chromatographic equipment are quite cumbersome and expensive. Meanwhile, they inevitably require to open the liquid container. Existing works also leverage RF signals, e.g., RFID [11] and UWB radar signals[9], to measure the liquid properties. The intuition is that RF signals traveling through or reflected by different liquids show different patterns of the signal parameter, e.g., the phase [37] or time-of-flight [9], which can be used for liquid detection. However, RF-based methods require specialized devices and cannot detect the liquids with metal containers since metals could affect the normal transmission and communication of the RF signals, which limits its usage scenarios. Considering the limitations of existing liquid detection solutions, we propose to detect liquid fraud in a cost-effective, non-intrusive, and ubiquitous manner.

5.3.1 *Acoustic-Based Liquid Fraud Detection*

In this case study, we aim to achieve a liquid fraud detection system using commodity acoustic devices, i.e., the speaker and microphone. The proposed system can detect the counterfeiting and adulterated liquids in real time, even without opening the liquid container. The speaker and microphone are clung to the surface of the liquid container on each side horizontally. The speaker emits the acoustic signal, and the microphone receives the acoustic signal traveling through the liquid. The insight comes from the fact that liquids with different components have different acoustic impedance, which determines the absorption of the acoustic signal [18]. Thus, the acoustic signal traveling through the liquid has the potential to distinguish the fake liquids from the authentic one due to their different patterns in absorbing the acoustic signal. In this case study, we adopt a hybrid way to combine model-based and data-driven approaches for liquid sensing using the acoustic signal. We first build a neural network model to detect fake liquids using the acoustic absorption pattern by the liquid. We further revise the loss function in the learning model to enhance the detection performance based on the effect of frequency-selective fading as we have introduced in Sect. 2.3.2.4.

5.3.1.1 Understanding the Acoustic Absorption and Transmission in Liquids

We employ the liquid's absorption of acoustic signal for liquid detection. The acoustic energy can be absorbed when the acoustic signal travels through the liquid. This is because the acoustic pressure facilitates the movement of liquid particles, resulting in internal frictions caused by the viscosity effect, which converts the acoustic energy into heat and induces the absorption of acoustic signal [18]. The

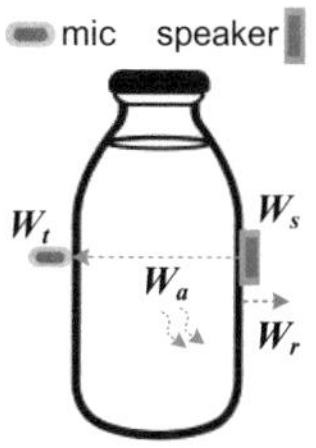

Fig. 5.70 Process of acoustic signal traveling through liquid

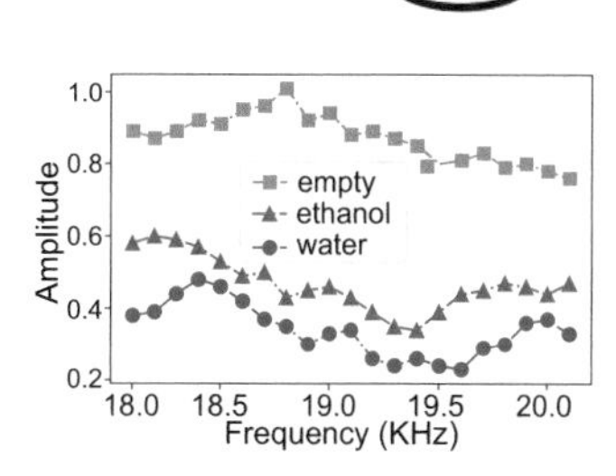

Fig. 5.71 Liquid absorption of acoustic signal in different medium

absorbed energy reaches its maximum when the acoustic frequency matches the liquid's natural frequency of vibration, i.e., the acoustic resonance phenomenon.[2] We model the process of transmitting the acoustic signal from the speaker on the right side of liquid to the microphone on the left side in Fig. 5.70. During this process, the acoustic signal sent by the speaker (W_s) first encounters the liquid container. Then, part of the signal is reflected by the container surface (W_r). Part of the signal is absorbed by the liquid and transformed to heat (W_a). Finally, part of the signal travels through the liquid and is received by the microphone (W_t). If we keep the sent signal W_s and container unchanged, the energy of the received signal (W_t) is mainly decided by how much signal is absorbed in the liquid, i.e., the W_a.

To show whether the acoustic signal can be affected by the liquid's absorption, we perform an experiment to compare the acoustic absorption without and with the water filled in a plastic bottle. We place one pair of speaker and microphone on two sides of the liquid container as shown in Fig. 5.70. Then, we transmit the acoustic signal with equal power on multiple frequencies, i.e., 18, 18.1, ..., 18.9, and 20 KHz. Then, we perform Fast Fourier Transformation (FFT) on the received signal and obtain the frequency-domain amplitude of each frequency. We keep all other settings unchanged during the experiment. As shown in Fig. 5.71, the amplitude of the empty bottle is higher than that of the bottle filled with water, showing that part of the sound energy is indeed absorbed by water.

The absorbed energy W_a is governed by the acoustic impedance (Z) of the liquid and is a function of frequency (f), i.e., $W_a(f) \sim Z_f$ [20]. The acoustic impedance is affected by the density of liquid (ρ) and the traveling speed (c) of acoustic signal in the medium, i.e., $Z = \rho \cdot c$. Since the density and sound speed are determined by

[2] The acoustic resonance phenomenon will rarely happen for our case since the resonant frequencies of liquids are around gigahertz-level, while the sound we transmit is in the 18–20 KHz frequency band.

liquid components, liquids with different components can result in different acoustic impedance. Thus the absorbed energy W_a varies accordingly. To investigate the effect of different liquid components on the absorption of acoustic signal among multiple frequencies, we conduct an experiment to compare the acoustic absorption for two different liquids. We prepare water (density: 1.0 g/cc, speed: 1482 m/s under 25 °C) and ethanol (density: 0.79 g/cc, speed: 1159 m/s under 25 °C). Then, we fill the same amount of water and ethanol in the same containers and remain other settings unchanged. As shown in Fig. 5.71, the amplitudes of all frequencies for the received acoustic signal traveling through ethanol are larger than that through water, which shows that more energy is absorbed by water due to its higher density and sound speed than those of ethanol. In addition, amplitudes of the received signal vary among different frequencies across different liquids since W_a is affected by the sound frequency as well. Therefore, the acoustic absorption and transmission curve (AATC), which is composed of amplitudes over multiple frequencies of the acoustic signal after being absorbed and transmitting through the liquid, can serve as a good feature to differentiate different liquids. We will introduce the design of W_s, W_t processing, and AATC extraction in the following content.

Feasibility of Using AATC for Liquid Detection

To investigate the feasibility of using the AATC for distinguishing different liquids and liquid fraud detection, we first conduct a set of preliminary experiments to observe the AATCs for: (1) different kinds of liquids; (2) random mixtures of one liquid with other fraudulent liquids; (3) mixtures of one liquid with different percentages of another fraudulent liquid. For (1), we select 3 kinds of alcohol products (liquor, ethanol, and isopropanol) and 3 kinds of cooking oil products (olive oil, canola oil, and soybean oil) in the market. For (2), we regard the ethanol and isopropanol as the fraudulent alcohol against the liquor and treat the canola oil and soybean oil as the fraudulent oil against the olive oil. Then, we randomly mix the liquor with ethanol and isopropanol, as well as mixing the olive oil with canola oil and soybean oil, respectively.

For (3), we mix the liquor and olive oil with different percentages of isopropanol (30 and 40%) and canola oil (20 and 30%), respectively. For each of the original and mixed liquids, we collect 3 traces of acoustic signal and extract the AATC from each trace. During the experiment, we use the same acoustic devices and container for all the liquids. We emit the acoustic signal with 21 frequencies ranging from 18 to 20 KHz with an interval of 100 Hz. Figure 5.72a–c depicts the AATCs of the liquor, ethanol, isopropanol, and their mixtures.[3] The AATCs of the olive oil, canola oil, soybean oil, and their mixtures are shown in Fig. 5.72d–f. From Fig. 5.72, we have the following observations: (1) The AATCs of the same liquid exhibit similar patterns and are stable across different times of measurements. (2) For different kinds of liquids, as shown in Fig. 5.72a d, the AATCs show distinct

[3] "etha" and "isop" in Fig. 5.72b, c are the abbreviations of the ethanol and isopropanol, respectively. (r) refers to the random mixture.

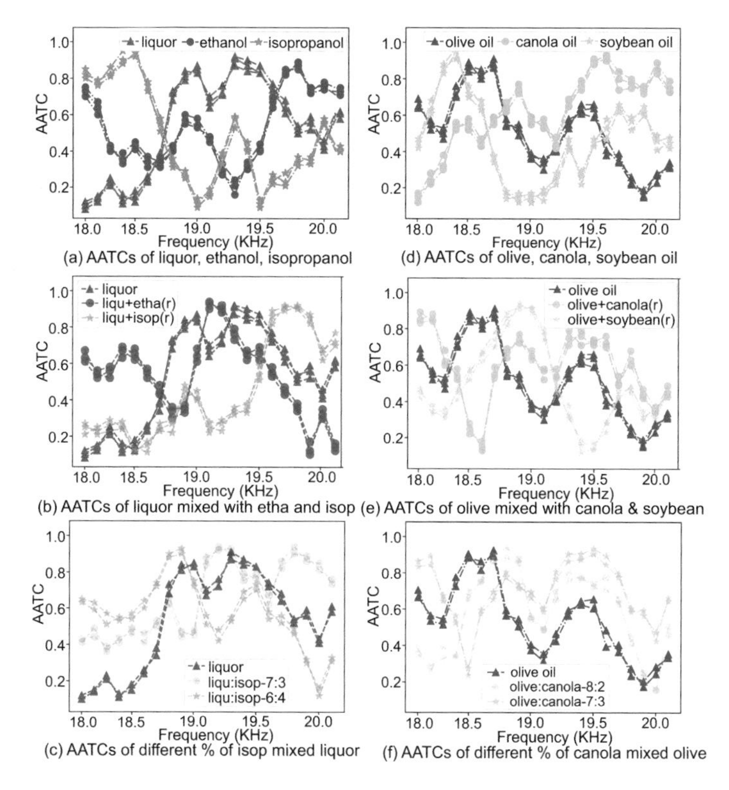

Fig. 5.72 AATCs of different authentic and adulterated liquids: (**a**–**c**) AATCs of different kinds and ratios of liquor and alcohol, (**b**) AATCs of different kinds and ratios of oil

patterns, indicating that the AATC is potential to distinguish different liquids. (3) As shown in Fig. 5.72b, e, the AATCs of the authentic liquids are distinct from those of the fake liquids mixed with different fraudulent liquids. Furthermore, as depicted in Fig. 5.72c, f, the AATCs of the authentic liquids compared with those of the fake liquids mixed with different percentages of the fraudulent liquid are different as well. This indicates that it is potential to use the AATC for detecting the fake liquids with different kinds and percentages of fraudulent liquids out of the authentic liquid.

Practical Factors for AATC Extraction

Although AATC is useful for liquid fraud detection, in practice, AATC extraction is vulnerable to multiple practical factors, which may significantly affect the liquid detection result. The first factor comes from the hardware diversity and imperfection of acoustic devices. The frequency responses of commodity speakers and microphones vary a lot across different frequencies, especially for the high-frequency band above 17 KHz [15]. The acoustic devices' frequency responses could affect the frequency-domain amplitudes of the received acoustic signal, resulting in inconsistent AATCs for the same liquid. To show the effect of different acoustic devices on AATC extraction, we use the same liquid but apply two different

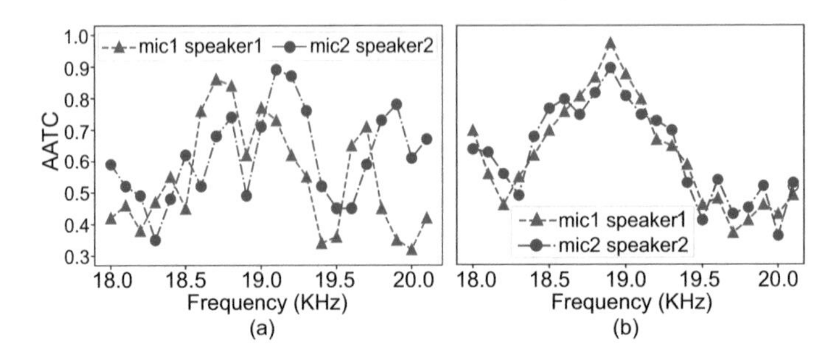

Fig. 5.73 Raw and calibrated AATCs of the same liquid using different speakers and microphones. (**a**) AATCs of different acoustic devices. (**b**) Calibrated AATCs in (**a**)

pairs of speakers and microphones to send and receive the signal. In Fig. 5.73a, the AATCs exhibit dissimilar patterns when using different acoustic devices for the same liquid. Hence, the AATC can be greatly affected by the frequency responses of the acoustic devices. In Sect. 5.3.1.4, we will introduce our proposed AATC calibration method to remove the effect of acoustic devices' frequency responses.

The second factor lies in the different relative positions between acoustic devices and liquid container. In practice, the position of acoustic devices relative to the container may be hard to keep the same when using the system at different times. The change of the relative device-container position can affect the AATC. This is because the acoustic signal traveling from different parts of the container can result in different multipath signals inside the container. Figure 5.74a shows the propagation paths of the acoustic signal in the liquid. The line-of-sight (LoS) signal remains unchanged when the acoustic devices are placed at different heights relative to the container. However, multipath signals reflected by the liquid and container (dashed lines) change along with different positions. Those changes result in variations of the received acoustic signal, making the AATCs measured from different positions vary for the same liquid. To show the effect of different positions on the AATC, we place the same acoustic devices at 5 different heights of the container. Then, we extract the AATCs for the olive oil at these 5 positions, as shown in Fig. 5.74c. The AATCs for the same liquid exhibit variations across the frequency band under different relative positions. These variations in the AATC could lead to misdetection of the same liquid. We perform preliminary experiments on detecting the authentic and fake olive oil with the same acoustic devices placed at different relative positions to the container. We use the authentic olive oil's AATCs collected at one position to train the one-class SVM model. Then, the AATCs of the authentic and fake olive oil collected at another 4 positions are used for testing. The detection accuracy decreases to 71.3% with errors mainly coming from inaccurately detecting the authentic olive oil as the fake one. An intuitive solution to promote the accuracy is to collect the acoustic signal from as many as relative device-container positions for the authentic liquid to train the anomaly detection model. However, it is labor-intensive or even impractical to collect such a large number of data. Thus,

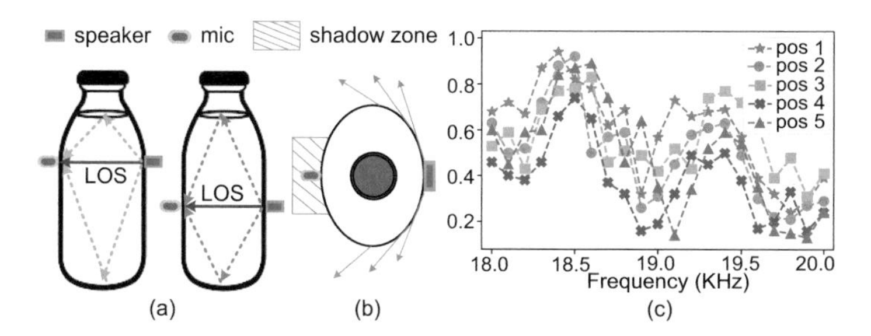

Fig. 5.74 Effect of relative device-container positions: (**a**) signals in liquid, (**b**) acoustic shadow zone, (**c**) AATCs of different positions

an alternative method is needed to deal with the insufficient training data. We will elaborate on our data augmentation method in Sect. 5.3.1.5.

5.3.1.2 Acoustic Signal Generation

The emitted acoustic signal $s(t)$ is designed as the sum of multiple sine waves with different frequencies, i.e., $s(t) = \sum_{i=1}^{n} A_i sin(2\pi f_i t)$, where A_i is the amplitude of each sine wave, f_i is the frequency, and n is the number of discrete frequencies. In our design, A_i is the same for all the sine waves. The discrete frequency f_i is within the frequency band of [18, 20 KHz]. The reason for choosing this frequency band lies in four aspects. First, the acoustic signal in this frequency band is inaudible to most people, which does not disturb the users when using the system. Second, frequencies of most background noises in the environment, as well as the human voice, are lower than 18 KHz [27]. Then, the noises in the environment are removed with a high-pass filter. Third, the acoustic signal's wavelength within such a frequency band is much smaller than the size of most containers, which can alleviate the sound diffraction effect. Finally, the upper bound of the frequency for most commodity speakers and microphones is 20 KHz. The interval I_f between every two discrete frequencies is equal, and I_f determines the granularity of AATC. In Sect. 5.3.1.7, we will discuss the effect of AATC granularity on the liquid detection performance. Finally, $s(t)$ is saved as a WAV file which is played by the speaker. Figure 5.75a shows the spectrum of $s(t)$ with 21 frequencies, i.e., $I_f = 100$ Hz.

5.3.1.3 Signal Pre-processing and Acoustic Feature Extraction

After emitting the acoustic signal from the speaker, the microphone receives the acoustic signal for 4 s with a sampling rate of 48 KHz. Then, a high-pass filter with a cutoff frequency of 18 KHz is applied on the received acoustic signal $r(t)$ to remove the background noises. Next, we perform FFT on the filtered signals. A Hamming window is applied on the filtered signal before FFT to reduce the frequency leakage.

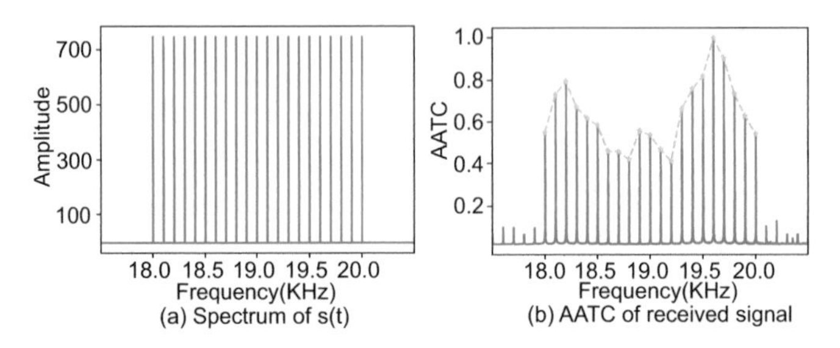

Fig. 5.75 (**a**) Spectrum of generated acoustic signal and (**b**) AATC extracted from the received signal

Then, we extract the frequency-domain amplitude at f_i as $R(f_i)$. Finally, $R(f_i)$ is divided by the corresponding amplitude $S(f_i)$ in the spectrum of $s(t)$ to obtain the AATC. AATC represents the ratio of the remaining acoustic signal's energy over the emitted acoustic signal's energy across multiple frequencies. In practice, the volume of the speaker and microphone may change. Thus, we normalize the AATC to the same scale of [0, 1] after AATC calibration. Figure 5.75b shows the normalized AATC, as denoted by the yellow dashed curve.

5.3.1.4 Tackling the Effect of Different Acoustic Devices

Modeling the Transmission of the Acoustic Signal from the Speaker, Liquid and, Its Container to the Microphone

The transmission of the acoustic signal in the whole system can be modeled as follows: (1) Frequency responses of speaker and microphone. The speaker and microphone are typical linear time-invariant (LTI) systems, which produce an output signal $y(t)$ from any input signal $x(t)$ subject to the constraints of linearity and time-invariance. The characteristic of a LTI system is described by its impulse response $h(t)$. Figure 5.76a shows the relationship among $x(t)$, $h(t)$, and $y(t)$ of the LTI system. The impulse responses of the speaker and microphone are denoted as $h_s(t)$ and $h_m(t)$, respectively, and the corresponding frequency responses are $H_s(f)$ and $H_m(f)$. (2) Acoustic signal transmission in the liquid and its container. When the acoustic signal travels through a medium, due to the reflection of the obstacles in the medium, there are multiple paths of the signal with different delays arriving at the receiver. The received signal can be modeled as a LTI system as well, which can be expressed as $y(t) = \sum_{i=1}^{N} a_i x(t - \tau_i) = h(t) * x(t)$, where $h(t)$ is the signal's channel impulse response in the medium, N is the number of paths, a_i and τ_i are the amplitude and time delay of each signal path, respectively. When the acoustic signal travels through the container, its channel impulse response $h_c(t)$ is mainly affected by the container's material and thickness. For the acoustic signal traveling through the liquid, i.e., $y_l(t) = \sum_{i=1}^{N_l} a_{l_i} x(t - \tau_{l_i}) = h_l(t) * x(t)$, a_{l_i} and τ_{l_i} in its channel

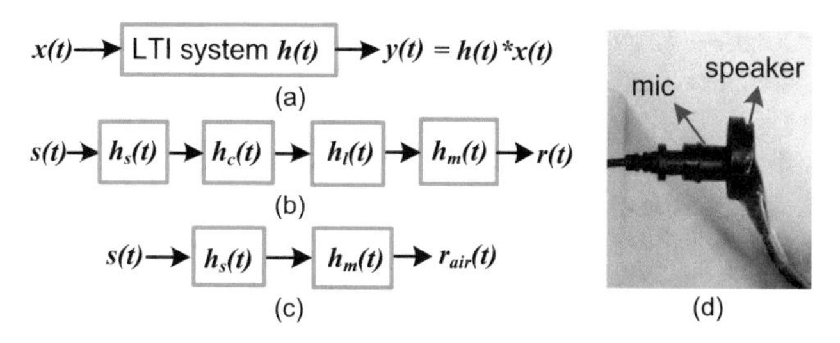

Fig. 5.76 Modeling of acoustic signal transmitting from speaker to microphone and the setting for AATC calibration. (**a**) Input, impulse response, output of LTI system. (**b**) Signal from speaker, container, liquid, to mic. (**c**) Signal from speaker, the air, to mic. (**d**) Reference setting

frequency response $h_l(t)$ contain the information about the liquid's absorption of the acoustic signal.

Finally, as modeled in Fig. 5.76b, the overall received acoustic signal $r(t)$ after the emitted signal $s(t)$ traveling through the cascade of the above four LTI systems can be expressed as $r(t) = s(t) * h_s(t) * h_c(t) * h_l(t) * h_m(t)$. By transforming $r(t)$ into the frequency domain, it becomes $R(f) = S(f) \cdot H_s(f) \cdot H_c(f) \cdot H_l(f) \cdot H_m(f)$, where $H_c(f)$ and $H_l(f)$ are the channel frequency responses in the container and liquid, respectively. When using different acoustic devices to measure the AATC for the same liquid and container, $S(f)$, $H_c(f)$, and $H_l(f)$ keep unchanged, while $H_s(f)$ and $H_m(t)$ are different, resulting in different $R(f)$ with inconsistent AATCs.

Calibrating the AATC Using the Reference Signal

To remove the effect of the acoustic devices' frequency responses, we design a reference signal which directly travels from the speaker and microphone without other medium between them. Specifically, at the initialization stage of the system, we place the speaker and microphone close to each other without space between them, as shown in Fig. 5.76d. Under this setting, the acoustic signal would directly travel from the speaker and microphone without other medium between them. Although there is still some air inside devices, the portion is quite small, so it can be neglected. Then, the measured reference signal $r_{ref}(t) = s(t) * h_s(t) * h_m(t)$. The corresponding frequency-domain representation becomes $R_{ref}(f) = S(f) \cdot H_s(f) \cdot H_m(f)$. Since we use the same acoustic devices to measure the reference signal and liquid's AATCs, dividing the reference signal by the signal traveling through the liquid in the frequency domain becomes the following equation.

$$
\begin{aligned}
\frac{R_{ref}(f)}{R_l(f)} &= \frac{S(f) \cdot H_s(f) \cdot H_m(f)}{S(f) \cdot H_s(f) \cdot H_c(f) \cdot H_l(f) \cdot H_m(f)} \\
&= \frac{1}{H_c(f)} \cdot H_l(f).
\end{aligned}
\tag{5.52}
$$

It shows that the calibrated signal is irrelevant to $H_s(f)$ and $H_m(f)$. In addition, $H_c(f)$ is a constant factor for the same type of liquids to be detected. Thus, the calibrated AATC is only affected by the liquid's frequency response $H_l(f)$, which can reflect the acoustic absorption of the liquid. Note that such a setting for calibration is a one-time setup before liquid detection, and the frequency response does not need to be calibrated again with the same speaker-microphone pair. Based on the above AATC calibration method, we calibrate the raw AATCs measured with different acoustic devices in Fig. 5.73a. The calibrated AATCs are shown in Fig. 5.73b. The calibrated AATCs when using different speakers or microphones exhibit similar patterns for the same liquid, which shows the effectiveness of our calibration method.

5.3.1.5 Tackling the Effect of Different Relative Device-Container Positions

Ideally, the collected AATCs should involve all the variations caused by different relative device-container positions to train the liquid detection model. However, manually collecting the AATCs from as many as possible positions is quite labor-intensive. In our work, we adopt a data augmentation technique which can automatically emulate the variations in AATCs caused by different relative device-container positions.

To find a proper method for AATC augmentation, we investigate the characteristics of the AATCs extracted from different relative device-container positions. We first choose one initial position at the center of the liquid container to place the speaker and microphone. Then, we move the speaker-microphone pair up, down, left, and right with 1 cm stepwise, as shown in Fig. 5.11. In sum, 10 different pairs of positions are selected for the speaker and microphone, as shown in Fig. 5.77. In Fig. 5.78, we depict the distribution of the AATCs extracted with the acoustic devices placed at 10 different positions relative to the container for the same liquid. For each position, 5 AATCs are extracted. Figure 5.78 shows that the extracted AATCs share similar distributions at different positions. We also measure the AATC's distribution for another 10 liquids and observe similar patterns. We further apply the equivalence test on the AATCs of different positions to check whether they follow the same distribution. The equivalence interval is set to the average difference among the AATCs collected from the same position, i.e., 0.03 in our experiment. The average p-value is 0.019 (threshold as 0.05), which rejects the hypothesis that the difference among the AATCs of different positions is larger than the equivalence interval. This indicates that the same liquid shares the same AATC distribution even at different relative device-container positions. As such, we can employ the generative model, which can generate new data following the same distribution of the input data with some variations, to augment the AATCs. In our work, we employ VAE for AATC augmentation because it can effectively augment more data based on a small amount of input data.

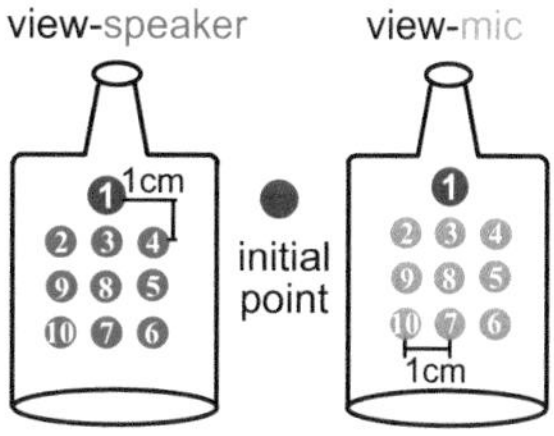

Fig. 5.77 Positions of speaker and microphone

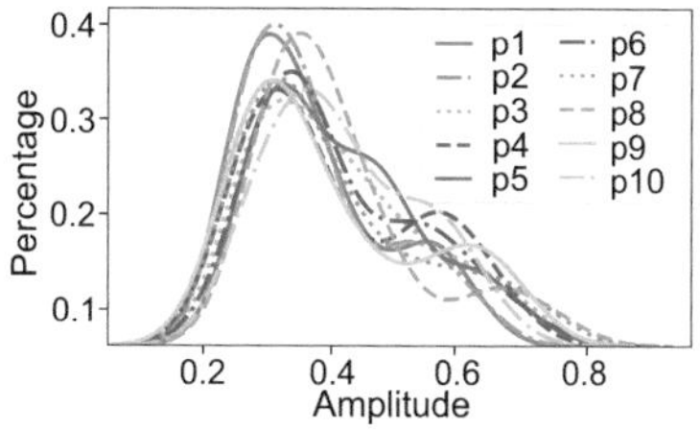

Fig. 5.78 Distribution of the same liquid's AATCs for devices at 10 positions

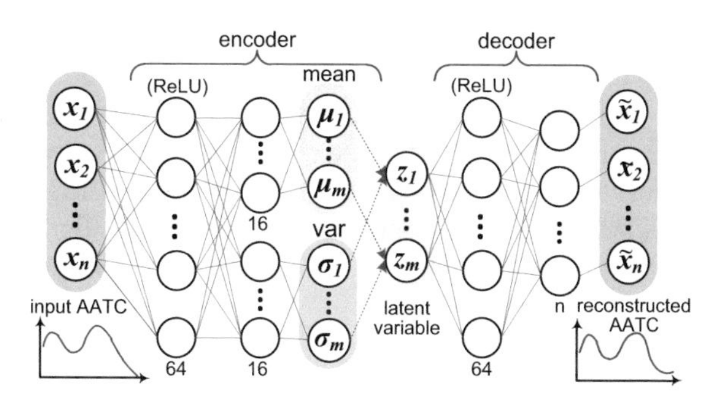

Fig. 5.79 Architecture of the VAE model which augments AATCs for different relative device-container positions

Figure 5.79 shows the VAE model for AATC augmentation. The input $x(n)$ is the vector of AATC, which is extracted from the manually collected acoustic signal. The output $\widetilde{x}(n)$ is the reconstructed AATC. The VAE model consists of an encoder whose target is to compress the input feature vector into a latent variable vector $z(m)$ and a decoder which decompresses the $z(m)$ to reconstruct the input. m is the length of the latent variable vector. Since the latent variable vector learns a representation with fewer dimensions than the input, m should be smaller than n. For our case with $n = 21$, $m < 21$. Meanwhile, considering that too few dimensions of z could lead to larger information loss, we empirically select $m = 16$, which achieves the highest accuracy when using the generated AATCs to train the model for liquid fraud detection.

To further enhance the performance of VAE for AATC augmentation, we add a regularizer in VAE's loss function based on a key observation about the AATCs of different relative device-container positions. We find that the AATCs on some frequencies experience larger variance than other frequencies at different positions. This is mainly due to the frequency-selective fading effect of the acoustic signal. In specific, the change of multipath acoustic signals caused by the position difference could strengthen or weaken the amplitude of the received acoustic signal with a

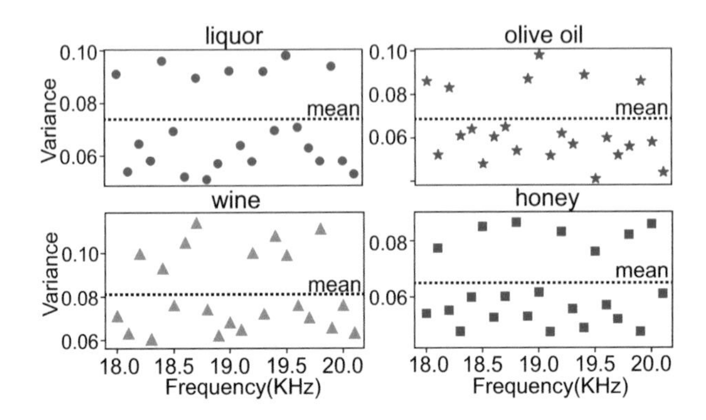

Fig. 5.80 Variances of the AATCs at different frequencies for different relative device-container positions among different liquids

larger degree on some frequencies. To show the frequency-selective fading effect on the AATCs of different positions, we depict the variances of AATCs over all the frequencies for the authentic liquor, olive oil, wine, and honey in Fig. 5.80. It shows that AATCs exhibit larger variances at several frequencies, i.e., the frequencies whose variances are above the mean of all the variances. This indicates that some frequencies are more sensitive to different positions. Recall that our purpose of using VAE is to generate AATCs that seem like being obtained under different device-container positions. To this end, we add a frequency-sensitive regularizer in the VAE's loss function $\mathcal{L}$ to enlarge the AATCs' variances for those sensitive frequencies as follows:

$$\mathcal{L} = \underbrace{E_{q_\theta(z|x_i)}[\log p_\phi(x|z)]}_{reconstruction\ loss} - \underbrace{KL(q_\theta(z|x)\|p(z))}_{KL\ divergence} - \underbrace{\|AATC(f_{sen}) - A\tilde{AT}C(f_{sen})\|_1}_{frequency-sensitive\ regularizer} \tag{5.53}$$

In Eq. (5.53), the first and second terms, which are the reconstruction loss between the input and generated AATCs and the Kullback-Leibler (KL) divergence, form the original VAE loss function. The third term is our added frequency-sensitive regularizer, where $AATC(f_{sel})$ and $A\tilde{AT}C(f_{sel})$ are the input and generated AATCs' values on the sensitive frequencies f_{sen}, respectively. To select f_{sen}, we first calculate the variances of the manually measured AATCs for each frequency and obtain the mean of all the variances. Then, the frequencies whose variances exceed the mean are selected as the f_{sen}. When training the VAE model, $\mathcal{L}$ is minimized to find the optimal weights in Eq. (5.53), meanwhile, the difference between the input and generated AATCs' values on those sensitive frequencies is enlarged. Finally, based on a certain number of manually measured AATCs, the VAE model will generate more AATCs for the authentic liquid, which are combined with the manually measured AATCs to train the liquid detection model.

5.3.1.6 Liquid Detection

Liquid Fraud Detection

To detect liquid fraud, intuitively, we can collect data from both authentic and fake liquids to train a binary classification model. However, in practice, it is difficult or impractical to acquire all kinds of fake liquids with various fraudulent components and adulteration ratios. Thus, we regard fake liquids as anomalies towards the authentic liquid and propose to build an anomaly detection model only using the AATCs of the authentic liquid.

We employ the VAE to build the anomaly detection model. The principle of using VAE for anomaly detection lies in the differences between the reconstruction losses of authentic and fake liquids. When training the VAE model using the authentic liquid's AATCs, the reconstruction loss between the input and generated AATCs is minimized. Then, VAE can learn how to generate new AATCs following the same distribution of the authentic liquid. When the testing input is the AATC of an authentic liquid, the reconstruction loss can be quite small. While if the testing input comes from a fake liquid since the AATCs of the authentic and fake liquids have different distributions, the reconstruction loss would be larger than that of the authentic liquid. Therefore, we can use the reconstruction loss of VAE to train the anomaly detection model. Specifically, we obtain all the reconstruction loss values when using the authentic liquid's AATCs to train the VAE model in Fig. 5.79. Then, we follow the three-sigma rule of thumb to select the threshold δ_t to detect the anomalies. The mean (μ_t) and standard deviation (σ_t) of all the losses are calculated. We compare the liquid fraud detection accuracy using $\mu_t+\sigma_t$, $\mu_t+2\sigma_t$, and $\mu_t+3\sigma_t$ as the δ_t, respectively. We δ_t set to $\mu_t + \sigma_t$ since it achieves the best accuracy. For an unknown liquid, we input its AATC to the VAE model. If the reconstruction loss is larger than δ_t, it is detected as the fake liquid, and vice versa.

Liquid Adulteration Ratio Recognition

Apart from liquid fraud detection, we find that the AATC has the potential to differentiate the adulterated liquids with different adulteration ratios according to the observations from Fig. 5.72c, f. For instance, the AATCs of mixing the liquor with isopropanol by the ratios of 7:3 and 6:4 show different patterns. This brings the opportunity for recognizing the liquid adulteration ratio. In practice, the liquids could be harmful to human health if the adulteration ratio exceeds a certain level. Therefore, it would be useful to recognize the liquid adulteration ratio using the AATC.

To achieve this, we first pre-define the interested adulteration ratios, e.g., 20, 30, and 40%, and collect the acoustic signal traveling through the adulterated liquids with different adulteration ratios. Next, AATC extraction, calibration, and augmentation are performed. Before inputting AATCs for training, we apply the Largest Margin Nearest Neighbor (LMNN) to map the AATC into a new space, so that the AATCs of the liquids with different adulteration ratios become more discriminative from each other. This is because LMNN can "pull" the AATCs of

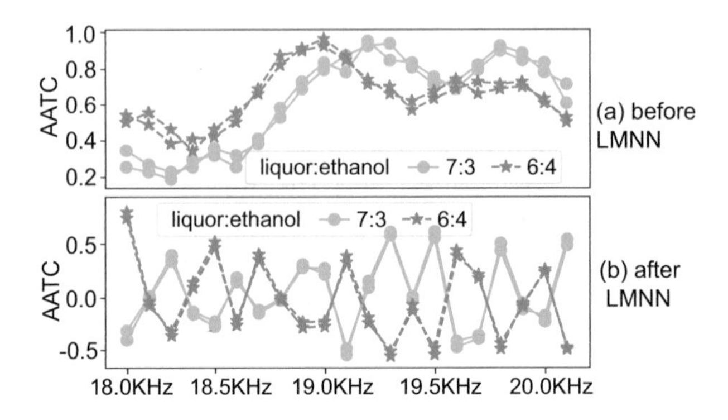

Fig. 5.81 AATCs before and after performing LMNN

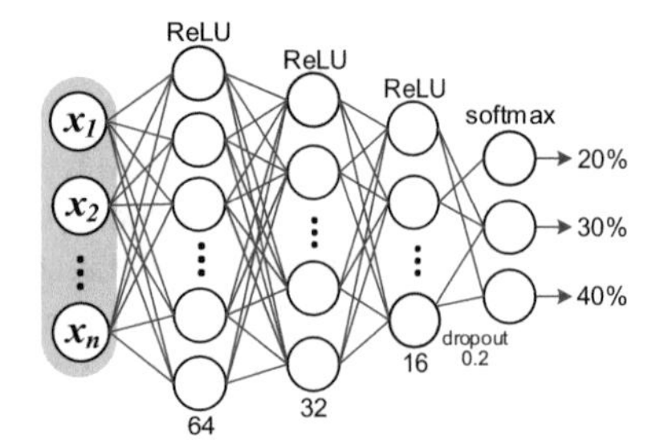

Fig. 5.82 MLP-based classification model for adulteration ratio recognition

the same class closer and meanwhile "push" the AATCs of different classes farther from each other. By doing this, LMNN can find a space in which AATCs of different adulteration ratios become larger while AATCs of the same ratio are narrowed. Figure 5.81 depicts the AATC before and after applying the LMNN for two ethanol-mixed liquor with close adulteration ratios. Compared with AATCs without LMNN, the transformed AATCs of the same liquid after LMNN are closer to each other, and AATCs of different liquids are of larger difference with each other. Finally, we apply the Multilayer Perceptron (MLP) neural network build the classification model, as shown in Fig. 5.82. For the fake liquid with an unknown adulteration ratio, its AATC is extracted and input to the classification model to obtain the adulteration ratio.

5.3.1.7 Evaluation

Hardware HearLiquid is implemented with commodity acoustic devices. The specifications of employed acoustic devices are listed in Table 5.5.

We use both commercial off-the-shelf external speaker-microphone pair and the bottom microphone in the smartphone as acoustic devices. As shown in Fig. 5.83, the acoustic devices are stuck to the two sides of the container surface using the adhesive type. The speaker and microphone are placed horizontally in the middle of the two sides of container. The acoustic devices can also be flexibly

Table 5.5 Specification of the acoustic devices

Device	Brand	Parameter	Frequency range
Speaker	iLouder	3 W, 8 Ω	700 Hz–20 KHz
	iLouder	2 W, 8 Ω	700 Hz–20 KHz
	HNDZ	3 W, 4 Ω	100 Hz–20 KHz
Mic	Sony D11	–	65 Hz–20 KHz
	BoYa M1	–	200 Hz–20 KHz
	Dayton imm-6	–	80 Hz–20 KHz

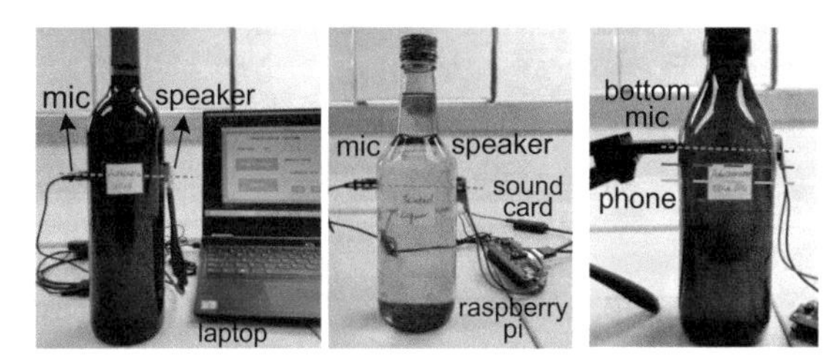

Fig. 5.83 System setup using external acoustic devices and smartphone

placed at different positions relative to the container, e.g., the red and blue dashed lines in the third subfigure in Fig. 5.83, for more convenient use of our system. External speaker-microphone pair is connected to the raspberry pi/laptop via a common sound card to send and receive the acoustic signal, respectively. The prices of external speakers and microphones are 5–15 US dollars. Apart from external acoustic devices, smartphones can also be used. For example, we can use the bottom speaker of the smartphone to send out the acoustic signal, as shown in Fig. 5.83.

Software We use Python to generate and process the acoustic signal. The generated acoustic signal is made of 21 sine waves whose frequencies range from 18 to 20 KHz with the same interval of 100 Hz. The signal is saved as a WAV file. The sampling rate of acoustic signal is 48 KHz, and the time duration for FFT is 4 s. Then the frequency resolution after FFT is 0.25 Hz, which is fine-grained enough to extract the amplitude on each desired integer frequency. The VAE and MLP models are trained via PyTorch on a server equipped with Intel(R) Xeon(R) CPU E5-2680 v2 and Nvidia GeForce RTX 2080 GPU with 32 GB memory. When training the model, we use the Adam optimizer and set the learning rate = 1e-4 and betas=(0.9, 0.999). The trained models are stored in the server for detecting the unknown liquid. To evaluate the performance of liquid fraud detection and liquid adulteration ratio recognition, the following metrics are used.

$$\begin{gathered} Accuracy = \frac{TP+TN}{TP+TN+FP+FN}, \\ Precision = \frac{TP}{TP+FP}, Recall = \frac{TP}{TP+FN}, \\ F1\,score = \frac{2 \cdot Precision \cdot Recall}{Precision + Recall} \end{gathered} \tag{5.54}$$

Table 5.6 List of authentic and fake liquids for different cases

Case	Liquids
Liquor fraud detection	Authentic: grey goose vodka (GGV, 40%)
	Fake: GGV + random e./m./i., stolichnaya vodka (40%)
Liquor ratio recognition	GGV + (30,35,40,45%) i.
Olive oil fraud detection	Authentic: colavita extra-virgin olive oil (Ceoo)
	Fake: ceoo + random c./p./s.
Olive oil ratio recognition	Ceoo + (25,35,70, 80%) c.
Wine fraud detection (different brands)	Authentic: torres mas la plana
	Fake: barefoot, penfolds, mirassou
Wine fraud detection (different grapes)	Authentic: barefoot (pinor noir)
	Fake: barefoot (shiraz, merlot, zinfandel)
Honey fraud detection	Authentic: comita manuka honey
	Fake: wildflower, lemon, longan flower honey
Honey MGO recognition	Comita manuka honey (MGO levels: 83+, 263+, 514+)

Liquid Data Collection We collect data from liquids of common liquid fraud cases in people's daily life, including liquor, extra-virgin olive oil, wine, and honey frauds, as listed in Table 5.6.

Authentic and tainted liquor: High-quality liquor is popular in many countries as daily drinks and gifts. Due to its high price, mixing authentic liquor with cheap and inedible alcohol for sale is the main method of liquor counterfeiting. Therefore, we prepare a high-quality authentic liquor product, the Grey Goose Vodka, and other kinds of alcohol, including ethanol (e.), methanol (m.), and isopropanol (i.), to mix with the authentic liquor as fake liquids. To show the system's ability to detect fake liquids with random mixtures of fraudulent liquids, we mix the authentic liquor with ethanol, methanol, and isopropanol, respectively. For each fraudulent alcohol, we make 3 bottles of fake liquids, which are obtained by randomly mixing the liquor with the corresponding alcohol. To evaluate the system for detecting the fake liquid which has the same alcohol level as the authentic liquor but with lower quality, we choose a cheaper vodka, Stolichnaya Vodka as the fake liquor. We also mix the authentic liquor with different percentages of the isopropanol, including 30, 35, 40, and 45%, to evaluate the performance of adulteration ratio recognition.

Wine fraud: Relabeling cheap wines to expensive ones is a common wine fraud. The expensive wines can be simply counterfeited by changing the wine label. In this case, we prepare an expensive wine, Torres Mas La Plana (grape: Cabernet Sauvignon), as the authentic wine, and three cheap wines, Barefoot California (grape: Merlot), Penfolds Koonunga Hill (grape: Shiraz), and Mirassou California (grape: Pinot Noir) with different grape types and brands as the fake liquids. In addition, to test whether the wine with different grape types can be detected, we prepare four wines of the same brand Barefoot California but different grape types, including Pinot Noir, Shiraz, Merlot, and Zinfandel. Pinot Noir is regarded as the authentic wine, and the other three types of grapes are treated as fake wines against Pinot Noir.

Honey fraud: Honey is the third most faked food in the world. The quality of honey varies a lot. The honey with a higher level of methylglyoxal (MGO) is much more expensive than ordinary honey. It is common that high-quality honey is replaced with poor one for sales. Thus, we prepare one high-quality honey, Comvita Manuka Honey with MGO, as the authentic honey, and select three cheaper honey, i.e., wildflower honey, lemon honey, and longan flower honey, as fake honey. In addition, to show the system's ability to recognize the Manuka honey with different MGO levels, we prepare three MGO levels Manuka honey (83+, 263+, and 514+).

Overall Performance First, we show the overall performance of the system on liquid fraud detection and adulteration ratio recognition. In Fig. 5.84, the accuracy, precision, recall, and F1 score are shown for all the liquid fraud cases. The accuracy of liquid fraud detection is around 92–96%. Specifically, for the liquor (liquor+**x**(r), **x**: ethanol, methanol, or isopropanol) and olive oil (olive+**x**(r), **x**: canola, soybean, or peanut oil) fraud detection, the average accuracy is approximately 95%, showing that the system can accurately detect the fake liquids with random mixtures of fraudulent liquids. Meanwhile, the system can detect the fake cheap liquor whose alcohol level is the same as the expensive authentic liquor with an accuracy of 92%. For wine fraud detection, the accuracy when detecting the fake wines whose brands and grape types are all different from the authentic wine is about 96%. While the accuracy drops a little to around 93% for detecting the fake wines whose brands are the same but with different grape types. The accuracy of detecting honey fraud is about 95%. We also prepare two bottles of honey with a similar MGO level (263+) but different brands to investigate whether our method can differentiate similar kinds of liquids but produced by different companies. We measure the AATCs from one

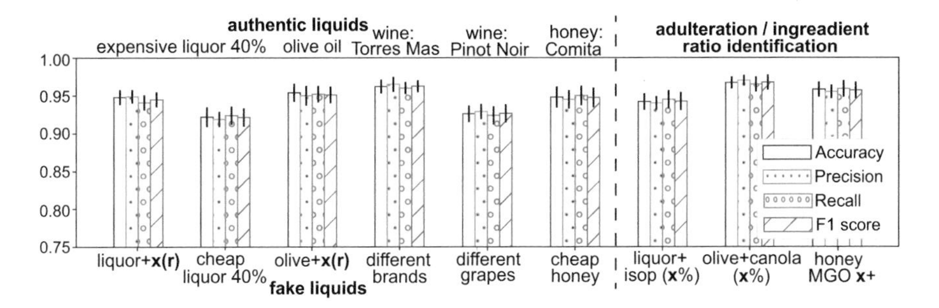

Fig. 5.84 Overall performance for liquid fraud detection and adulteration ratio recognition for different liquid fraud cases

bottle of honey (regarded as authentic honey) and train the liquid fraud detection model and use the AATCs collected from the other bottle of honey (regarded as fake honey) to test the model. The results show that 89.3% of the testing samples are accurately detected as fake honey. This is because, although having the same MGO level, they are still different in other ingredients, e.g., the amount of carbohydrate and sugar is different. This indicates that the components of the two types of honey still have some differences so that their absorption of the acoustic signal would be distinctive.[4]

For liquid adulteration ratio recognition, we train the ratio classification models for the isopropanol-mixed liquor and canola-mixed olive oil with 4 different adulteration ratios, respectively. As shown in the right part of Fig. 5.84, the accuracy for recognizing the adulteration ratio of liquor with the ratio difference of 5% is around 94%, and the recognition accuracy of the olive oil adulteration ratio with 10% interval can reach about 97%. Besides, the Manuka honey with different levels of MGO can be recognized with an accuracy of 95%.

Performance with Different Acoustic Devices In this evaluation, we show the performance of the system for liquid fraud detection using different acoustic devices. First, we use two different sets of acoustic devices, including external speaker and microphone, external speaker and bottom microphone of the smartphone, to train and test the liquid fraud detection model, respectively. When using both external speaker and microphone, the accuracy and F1 score are 0.956 and 0.954, respectively. When using the external speaker and the smartphone's bottom microphone, the accuracy and F1 score are 0.935 and 0.934, respectively. Second, we use different external speakers and microphones to evaluate our calibration method. We use 3 speakers (S1, S2, S3) and 2 microphones (M1, M2) to evaluate our proposed AATC calibration method for removing the effect of hardware diversity. S1 and S2 are from the same brand (B1) but with different specifications, while S3 is from another brand (B2). M1 and M2 are from different brands B3 and B4, respectively. In the experiment, we first use speaker B1-S1 and microphone B3-M1 to collect both the training and the first testing data. Then, to evaluate the detection accuracy of using different devices from the same brand, we keep the microphone B3-M1 while using another speaker B1-S2 to collect the second testing data. Furthermore, we use speaker B2-S2 and microphone B4-M2 to collect the third testing data to evaluate the performance of liquid fraud detection using different brands' devices. The results are shown in Fig. 5.85. Comparing with the accuracy of using the training and testing data from the same acoustic devices, the accuracy of using different acoustic devices with the same brand still exceeds 92% with only 3–4% decrease of accuracy, which shows the effectiveness of our AATC calibration method. The accuracy further decreases slightly to about 89–91% when

[4] We find our method does not work when the honey experiences crystallization because the honey components change in this process.

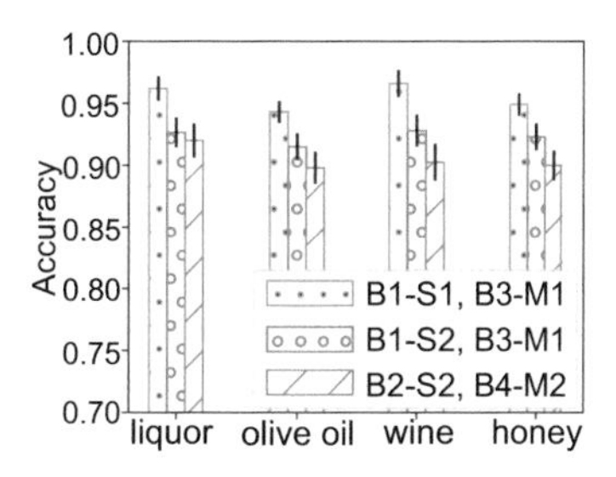

Fig. 5.85 Accuracy of liquid fraud detection (different acoustic devices)

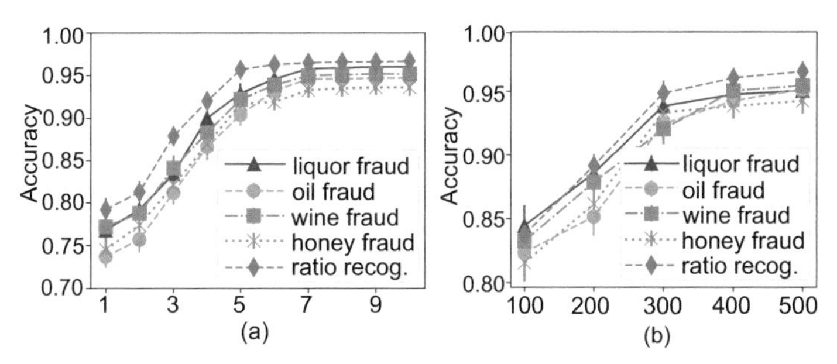

Fig. 5.86 Effect of the number of positions to collect the AATCs for training the VAE model and the number of augmented AATCs for liquid detection. (**a**) number of positions. (**b**) number of generated AATCs

using devices from different brands because the frequency responses of different brand's devices have larger deviations of the measured AATCs.

Impact of AATC Augmentation When using VAE, two factors can affect the liquid detection performance. The first factor is the number of relative device-container positions to collect the AATCs for training the VAE model. The second factor is the number of generated AATCs from VAE for training the anomaly detection and ratio recognition models. In this evaluation, we test the system performance for each factor.

First, we investigate the effect of different numbers of positions to collect the AATCs for training the VAE model. We choose 10 different positions and use different numbers of them (from 1 to 10) to collect AATCs and train the VAE model. Then, we ask volunteers to randomly collect AATCs from another 10 positions to test the model. We guarantee that the positions for training and testing do not overlap. At each position, 10 AATCs are collected. The AATCs generated from the VAE model are mixed with the manually collected AATCs to train the liquid detection models. We fix the number of generated AATCs from VAE to 400. The accuracy of using the AATCs from different numbers of positions is shown in Fig. 5.86a.

With more positions' AATCs to train the VAE model, the testing accuracy gradually increases. This is because the VAE model can learn more patterns from more device-container positions. In our experiment, for liquid fraud detection, the accuracy does not improve obviously after the number of trained positions exceeds 7. Therefore, we only need to collect the training AATCs from 7 positions (i.e., 70 AATCs), and the AATCs collected from other positions can be accurately detected. For ratio recognition, the AATCs from 5 different positions (i.e., 50 AATCs) are collected from each ratio of liquid to train the classification model, which already achieves an average classification accuracy of around 96%.

Second, we investigate the effect of different numbers of generated AATCs from VAE. We fixed the number of manually collected AATCs to 70 (for anomaly detection) and 50 (for ratio recognition) while using 100, 200, 300, 400, and 500 generated AATCs, which are combined with manually collected AATCs, to train liquid detection models. As shown in Fig. 5.86b, the accuracy increases with more number of generated AATCs. When the number of generated AATCs reaches 300, the average accuracy for all cases exceeds 90%. As the number increases to 400, the average accuracy exceeds 94%. When the number is above 400, our system shows no pronounced improvement. Thus, we generate 400 AATCs from the VAE to augment the training data.

References

1. Adib F et al (2015) Smart homes that monitor breathing and heart rate. In: Proceedings of the 33rd annual ACM conference on human factors in computing systems, pp 837–846
2. Ali Tayaranian Hosseini SM, Amindavar H (2017) UWB radar signal processing in measurement of heartbeat features. In: IEEE international conference on acoustics, speech and signal processing, pp 1004–1007
3. BBC News, Sleep position gives personality clue. http://news.bbc.co.uk/1/hi/health/3112170.stm
4. Bechbache RR, Duffin J (1977) The entrainment of breathing frequency by exercise rhythm. J Physiol 272(3):553–561
5. Benchetrit G (2000) Breathing pattern in humans: diversity and individuality. Respir. Physiol. 122(2–3):123–129
6. Bernasconi P, Kohi J (1993) Analysis of co-ordination between breathing and exercise rhythms in man. J Physiol 471(1):693–706
7. Bramble DM, Carrier DR (1983) Running and breathing in mammals. Science 219(4582):251–256
8. Bu Y et al (2020) RF-dial: rigid motion tracking and touch gesture detection for interaction via RFID tags. IEEE Trans Mobile Comput 21:1061–1080
9. Dhekne A et al (2018) Liquid: a wireless liquid identifier. In: Proceedings of the 16th annual international conference on mobile systems, applications, and services, pp 442–454
10. Ding H et al (2015) Femo: a platform for free-weight exercise monitoring with RFIDs. In: Proceedings of the 13th ACM conference on embedded networked sensor systems, pp 141–154
11. Ha U et al (2020) Food and liquid sensing in practical environments using RFIDs. In 17th USENIX symposium on networked systems design and implementation, pp 1083–1100
12. Hao T (2015) RunBuddy: a smartphone system for running rhythm monitoring. In: Proceedings of the ACM international joint conference on pervasive and ubiquitous computing, pp 133–144

13. Jha SN et al (2016) Detection of adulterants and contaminants in liquid foodsa review. Crit Rev Food Sci Nutrition 56(10):1662–1684
14. Ji T, Pachi A (2005) Frequency and velocity of people walking. Struct Eng 84(3):36–40
15. Lee H et al (2015) Chirp signal-based aerial acoustic communication for smart devices. In: IEEE conference on computer communications, pp 2407–2415
16. Liu X et al (2014) Wi-sleep: contactless sleep monitoring via wifi signals. In IEEE real-time systems symposium, pp 346–355
17. Liu J et al (2015) Tracking vital signs during sleep leveraging off-the-shelf wifi. In: Proceedings of the 16th ACM international symposium on mobile Ad Hoc networking and computing, pp 267–276
18. Markham JJ, Beyer RT, Lindsay RB (1951) Absorption of sound in fluids. Rev Mod Phys 23(4):353
19. McDermott WJ et al (2003) Running training and adaptive strategies of locomotor-respiratory coordination. Eur J Appl Physiol 89(5):435–444
20. Morse PM et al (1940) Acoustic impedance and sound absorption. J Acoust Soc Am 12(2):217–227
21. Nikitin PV et al (2010) Phase based spatial identification of UHF RFID tags. In: IEEE international conference on RFID, pp 102–109
22. Oppenheim AV et al (1989) Sampling of continuous-time signals. Discrete-time signal processing, vol 879. Prentice Hall, Hoboken
23. Persegol L et al (1991) Evidence for the entrainment of breathing by locomotor pattern in human. J Physiol 85(1):38–43
24. Piechowicz J (2011) Sound wave diffraction at the edge of a sound barrier. Acta Physica Polonica, A. 119, pp 1040–1045
25. Recinto C et al (2017) Effects of nasal or oral breathing on anaerobic power output and metabolic responses. Int J Exercise Sci 10(4):506
26. Richard GL (2004) Understanding digital signal processing, 3/E. Pearson Education India, Chennai
27. Rosen S, Howell P (2011) Signals and systems for speech and hearing, vol 29. Emerald, Bingley
28. Scholkmann F et al (2012) An efficient algorithm for automatic peak detection in noisy periodic and quasi-periodic signals. Algorithms 5(4):588–603
29. Scott Kelso JA (1995) Dynamic patterns: the self-organization of brain and behavior. Lecture notes in complex systems. MIT Press, Cambridge
30. Shafiq G, Veluvolu KC (2014) Surface chest motion decomposition for cardiovascular monitoring. Sci Rep 4(1):1–9
31. Wang Y, Zheng Y (2018) Modeling RFID signal reflection for contact-free activity recognition. Proc ACM Interact Mobile Wearable Ubiq Technol 2(4):1–22
32. Wang Y, Zheng Y (2019) TagBreathe: monitor breathing with commodity RFID systems. IEEE Trans Mobile Comput 19(4):969–981
33. Wang H et al (2016) Human respiration detection with commodity wifi devices: do user location and body orientation matter? In: Proceedings of the 2016 ACM international joint conference on pervasive and ubiquitous computing, pp 25–36
34. Wang X et al (2017) PhaseBeat: exploiting CSI phase data for vital sign monitoring with commodity WiFi devices. In: IEEE 37th international conference on distributed computing systems, pp 1230–1239
35. Wang X et al (2017) TensorBeat: tensor decomposition for monitoring multiperson breathing beats with commodity WiFi. ACM Trans Intell Syst Technol 9(1):1–27
36. Xie Y et al (2018) Precise power delay profiling with commodity Wi-Fi. IEEE Trans Mobile Comput 18(6):1342–1355
37. Xie B et al (2019) Tagtag: material sensing with commodity RFID. In: Proceedings of the 17th conference on embedded networked sensor systems, pp 338–350
38. Yang C et al (2014) AutoTag: recurrent variational autoencoder for unsupervised apnea detection with RFID tags. In: IEEE global communications conference, pp 1–7

39. Yang L et al (2014) Tagoram: real-time tracking of mobile RFID tags to high precision using COTS devices. In: Proceedings of the 20th annual international conference on Mobile computing and networking, pp 237–248
40. Zhang L et al (2018) Wi-run: multi-runner step estimation using commodity Wi-Fi. In: 15th annual IEEE international conference on sensing, communication, and networking, pp 1–9
41. Zhao R et al. (2018) CRH: a contactless respiration and heartbeat monitoring system with COTS RFID tags. In: 15th annual IEEE international conference on sensing, communication, and networking, pp 1–9
42. Zhu J et al (2014) RSSI based bluetooth low energy indoor positioning. In: International conference on indoor positioning and indoor navigation, pp 526–533

Chapter 6
Conclusion

6.1 Research Summary

Wireless sensing had appeared around a hundred years ago when Radar technologies were developed for military use. At that time, wireless communication is also far from people's daily lives. It is quite unbelievable that we only spend several decades bringing wireless communication for everyday use. Researchers with sharp minds find opportunities in the pervasive wireless signals and enable the presence of wireless sensing technology. We have witnessed the application of wireless sensing spreading into human beings, objects, environment sensing, and so on. In recent years, many emerging start-ups are converting wireless sensing techniques into commercial products, e.g., Emerald Innovations was founded by MIT faculty and researchers, Origin Wireless led by Ray Liu, and AIWise in China. We believe wireless sensing will grow faster in the future decade.

We have two critical observations based on the past decade's research in wireless sensing. First, the design of wireless hardware and signal is the core of wireless sensing. The sensing capability is mostly determined by the hardware and signal design. Many existing wireless sensing systems employ commercial off-the-shelf wireless devices, which suffer from many hardware imperfections. Meanwhile, the wireless signal cannot be configurable as well. For example, the Intel 5300 WiFi network interface card, the ImpinJ RFID reader, and the speaker and microphone embedded in the smartphone have been used as representative wireless devices to achieve sensing purposes. Employing commercial off-the-shelf devices can reduce the system cost. However, only employing those devices would limit the application scenarios and sensing performance. Therefore, some researchers have started to design their own wireless devices according to their expected sensing capability.

Second, a well-design wireless sensing platform requires researchers from different backgrounds, including electronic engineering, communication engineering, and computer science, to cooperate with other subjects. They need to work together to provide an end-to-end system with a holistic design of wireless signal

J. Cao, Y. Yang, *Wireless Sensing*, Wireless Networks,
https://doi.org/10.1007/978-3-031-08345-7_6

and hardware, signal processing, and function and interface implementation. For specific applications, it also requires knowledge from domain experts in the related applications. In this way, we can try to deal with many practical issues and remove as many constraints from the whole hardware and software design for deploying wireless sensing into wide use.

6.2 The Future

Wireless sensing aims to achieve seamless sensing by using the ubiquitous wireless signal around us. However, existing wireless signals are mainly designed for communication purposes. Therefore, reusing the signal for sensing purposes would disturb the communication functions. While, if we separately divide part of wireless resources to achieve sensing, the limited resources for communication become more severe. Therefore, a new paradigm, called integrated sensing and communication (ISAC), is proposed. The aim of ISAC is to jointly optimize the sensing and communication operations on a single hardware platform. ISAC is a promising technology to enable wireless sensing in a more friendly way.

We also call for more cooperation between academia and industry in the future. Existing wireless sensing systems are mainly tested in labs and simulated environments. The lack of large-scale testing impedes a deep understanding of practical deployment issues in the wild. To this end, academia needs support from the industry which can provide real-deployment scenarios and share their knowledge and experiences about user requirements for the sensing service. In fact, some problems and application scenarios proposed by researchers may not be a critical issue in practice. While, some of the fundamental problems can be overlooked by researchers due to the limited experience about actual demands. Therefore, the communication and collaboration between academia and industry should be further strengthened.

Last but not least, it would be quite meaningful to build an open community for wireless sensing. AI technologies may not achieve such a fast development without every researcher's open community and devotion. We notice a growing number of public events organized by known researchers to share their experiences in wireless sensing. We hope the wireless sensing community can be more open and absorb people from various backgrounds in the future.

Printed in the United States
by Baker & Taylor Publisher Services